BRITISH GEOLOGICAL SURVEY

British Regional Geology

Bristol and Gloucester region

THIRD EDITION

By G W Green, MA

Based on previous editions by
G A Kellaway, DSc, F B A Welch, BSC, PhD
and R Crookall, DSc, PhD

LONDON   HER MAJESTY'S STATIONERY OFFICE   1992

HER MAJESTY'S STATIONERY OFFICE

HMSO publications are available from:

HMSO Publications Centre
(Mail, fax and telephone orders only)
PO Box 276, London SW8 5DT
Telephone orders 071-873 9090
General enquiries 071-873 0011
(Queuing system in operation for both
  numbers)
Fax orders 071-873 8200

HMSO Bookshops
49 High Holborn, London WC1V 6HB
(Counter service only)
  071-873 0011   Fax 071-873 8200
258 Broad Street, Birmingham B1 2HE
  021-643 3740   Fax 021-643 6510
Southey House, 33 Wine Street, Bristol
  BS1 2BQ   0272-264306
  Fax 0272 294515
9–21 Princess Street, Manchester
  60 8AS   061-834 7201
  Fax 061-833 0634
16 Arthur Street, Belfast, BT1 4GD
  0232 238451   Fax 0232 235401
71 Lothian Road, Edinburgh
  EH3 9AZ   031-228 4181
  Fax 031-229 2734

HMSO's Accredited Agents
(see Yellow Pages)

And through good booksellers

BRITISH GEOLOGICAL SURVEY

Keyworth, Nottinghamshire NG12 5GG
0602-363100

Murchison House, West Mains Road,
Edinburgh EH9 3LA   031-667 1000

London Information Office, Natural
History Museum Earth Galleries,
Exhibition Road, London SW7 2DE
  071-589 4090

The full range of Survey publications is
available through the Sales Desks at Key-
worth and at Murchison House,
Edinburgh, and in the BGS London
Information Office in the Natural History
Museum Earth Galleries. The adjacent
bookshop stocks the more popular books
for sale over the counter. Most BGS
books and reports are listed in HMSO's
Sectional List 45, and can be bought from
HMSO and through HMSO agents and
retailers. Maps are listed in the BGS Map
Catalogue, and can be bought from
Ordnance Survey agents as well as from
BGS.

*The British Geological Survey carries out the
geological survey of Great Britain and Northern
Ireland (the latter as an agency service for the
government of Northern Ireland), and of the
surrounding continental shelf, as well as its
basic research projects. It also undertakes
programmes of British technical aid in geology in
developing countries as arranged by the Overseas
Development Administration.*

*The British Geological Survey is a component
body of the Natural Environment Research
Council.*

Maps and diagrams in this book use
topography based on Ordnance Survey
mapping

Contents

Figures   **Page**

Tables

Foreword to the Third Edition

The second edition of the Bristol and Gloucester regional guide was produced in 1948, but since that time much new information has been collected and there was a need to produce this third edition to incorporate the information derived from mapping carried out by the British Geological Survey and exploration undertaken by the coal, petroleum and water industries.

The Bristol and Gloucester district is one of the geologically most varied parts of Britain. This in turn provides the scenic variety for which the area is renowned, including the deep gorges of Cheddar Gorge and the Wye Valley, prominent escarpments such as those of the Cotswolds and the Forest of Dean, and the wide valleys flanking the Thames and the Severn. The area is also famous for its caves, such as Wookey Hole. It is the combination of a congenial environment and mineral resources which has attracted people to the Bristol and Gloucester district since the days of early man.

The geology of the district, as outlined in this volume, provides a fascinating insight into the history and evolution of the area, extending back more than 500 million years. The region contains some of the oldest rocks exposed in southern England whilst at the other end of the geological timescale, because the region was located near the edge of the massive ice sheets which at times covered much of Britain, it provides us with a dramatic picture of changing environments over the last two million years.

Elucidating the geological history is essential if we are to understand the controls on the distribution of mineral resources in the region. These resources include lime and aggregate, which are extracted in large quantities, and resources which are not presently mined, such as coal, evaporites, ironstone, clay, lead, (worked from Roman times), zinc and oil shales. Groundwater is one of the most important and widely used resources in the region; the geological investigations undertaken by the BGS in the region will help to conserve this important resource for present and future generations. Bath and other thermal waters have been the subject of recent investigations which have shown that their composition (and temperature) are closely related to their underlying geology.

The first two editions of this regional guide were very popular and I am confident that the third edition will be equally popular with geologists, planners, environmentalists, explorers, tourists and all those interested in the geological development of this beautiful area and its conservation.

British Geological Survey
Keyworth
Nottingham NG12 5GG

1 March 1992

Peter J Cook, DSc
Director

x

ACKNOWLEDGEMENTS

The author wishes to acknowledge the great help he has received from numerous colleagues on the Survey in writing this account; in particular to Dr P M Allen and Mr B J Williams for editing the text, and to members of the palaeontological and petrographical departments for evaluating appropriate sections. Special mention must also be made of Mr R J Wyatt, not only for checking the entire work and seeing it through the press but also for supplying valuable help with Chapter 12 (Great Oolite Group). The figures were prepared by Mr J W Arbon. Thanks are due to the Council of the Geological Society of London for permission to use and adapt Figure 15 (from the Journal of the Geological Society, 1983), and to Messrs Blackie for Figures 32 and 33.

Table 1 Main stratigraphical subdivisions present at outcrop in the Bristol–Gloucester region (not to scale). The ages shown refer to the worldwide system limits and only correspond to the base of the rock units where the local succession is complete.

Chronostratigraphical (time) units System	Series	Lithostratigraphical (rock) units	Age (10^6 years)
Quaternary	Holocene	Alluvium, peat, terrace deposits, raised-beach deposits, marine sands, head deposits, cave deposits, glacial deposits	10 000 years
	Pleistocene		about 2
Cretaceous		Chalk	
		Upper Greensand	
		Gault	130
Jurassic	Upper Jurassic	Oxford Clay and Kellaways Beds	
	Middle Jurassic	Great Oolite Group	
		Inferior Oolite Group	
	Lower Jurassic	Upper Lias	
		Middle Lias	
		Lower Lias	205
Triassic		Penarth Group	
		Mercia Mudstone Group	
		Sherwood Sandstone Group	250
?Permian		Bridgnorth Sandstone (Midlands)	
		unnamed sandstones (South-west)	290
Carboniferous	?Stephanian	Coal Measures	
	Cantabrian		
	Westphalian		
	Namurian	Quartzitic Sandstone Group	
	Dinantian	Carboniferous Limestone	365
Devonian	Upper Devonian	Upper Old Red Sandstone	
	Lower Devonian	Lower Old Red Sandstone	400
Silurian	Přidolí	Thornbury Beds Downton Castle Sandstone	
	Ludlow	Whitcliffe Beds, Leintwardine Beds, Bringewood Beds, Elton Beds	
	Wenlock	Brinkmarsh Beds	
	Upper Llandovery	Tortworth Beds	
		Damery Beds	418
Cambrian	Tremadoc	Micklewood Beds	
		Breadstone Shales	>475

1   Introduction

The area dealt with in this book comprises the Cotswolds and the Severn Estuary region, and includes the greater part of the counties of Avon, Gloucestershire and Somerset (excluding west Somerset); also, for geological continuity, small parts of the counties of Gwent, Herefordshire, Worcestershire, Wiltshire and Dorset.

Geologically speaking, it is one of the most varied districts of Britain, for, with the exception of the Ordovician and possibly the Permian, there is exposed at the surface every geological system from the Cambrian to the Cretaceous (Table 1).

The geological map (Figure 1) shows that the central part of the region is occupied by a triangular area of Palaeozoic rocks (concealed in many places by a thin covering of Mesozoic strata) extending from the Forest of Dean to the Mendips. To the south and north-east of this Palaeozoic triangle lie unbroken stretches of Mesozoic sedimentary rocks.

PHYSIOGRAPHY

The distribution of high and low ground (Figure 2) is related both to the nature of the underlying rocks and to the denudation to which they have been subjected. In general, the Palaeozoic rocks, by reason of their relative hardness, give rise to areas of moderately high relief. In parts of south Gloucestershire and Avon, however, the outcrops of the Cambrian, Silurian, Old Red Sandstone and Carboniferous beds are characterised by low undulating country. This apparent anomaly is due to the removal by erosion of the Mesozoic cover from these older strata, revealing part of an ancient erosion surface which had been reduced to very low relief before it was covered by sediments in Mesozoic times.

In the south-west and south of Bristol, Palaeozoic rocks form areas of moderate relief lying between 100 and 200 m above sea level. These comprise the Long Ashton to Clevedon ridge, which is separated by the Flax Bourton valley from the dome-like mass of Broadfield Down, with Dundry Hill, formed of Mesozoic rocks, on the north-east. This mass is limited to the south and east by the valleys of the Yeo and Chew rivers. Apart from the inlier of Cannington Park, near Bridgwater, the most southerly outcrop of Palaeozoic rocks in the district forms the elevated , undulating plateau of the Mendips, which extends from Frome on the east to Brean Down on the west, and continues out to sea in the rocky islands of Flat Holm and Steep Holm. Three hill ridges in the western part of the Mendips reach a height of just over 300 m.

In contrast with this, there lies to the south of the Mendips an expanse of alluvial flats (Plate 12) concealing extensive depressions scoured out of soft Mesozoic sediments. The dead flat of these moors or 'levels' is broken by the low ridge of the Polden Hills, Brent Knoll and the rising ground above Wedmore.

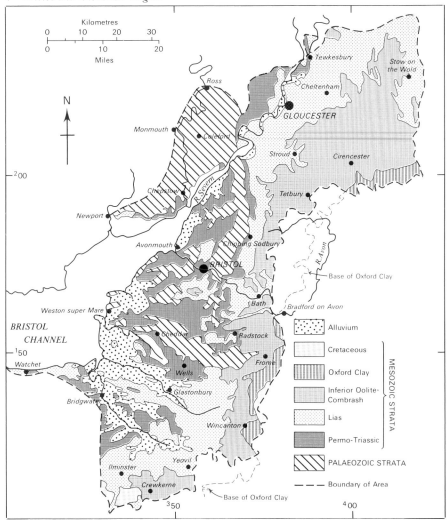

Figure 1   Sketch map of the geology of the region.

Running roughly parallel with the eastern margin of the district, comparatively hard Jurassic limestones form the mass of the Cotswolds, whose indented, wall-like scarp, overlooking the Vale of the Severn, runs from Chipping Campden to Bath. On Cleeve Common, near Cheltenham, the Cotswolds reach their maximum height of 330 m. Southwards to Bath the general height of the range gradually decreases, the hills around Bath attaining an altitude of 180 to 240 m. To the south of Bath, the Mesozoic strata form a tract of ground which, though seldom exceeding 175 m in general height, is very varied in relief, being deeply dissected by the River Avon and its tributaries, notably the River Frome and the Cam and Wellow brooks. This region merges into the eastern part of the Mendips, where the Mesozoic rocks form a thin intermittent covering to the Palaeozoic strata.

Between Bath and the Mendips the clay interval between the Jurassic limestones becomes more marked because of the replacement of the Great Oolite by the

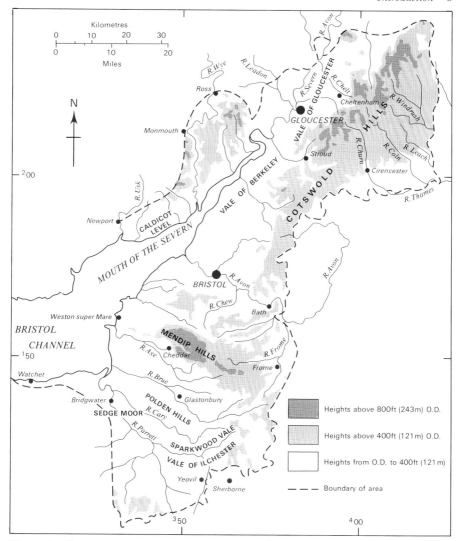

Figure 2 Sketch map of the physical features of the region.

Frome Clay. Two roughly parallel ridges, separated from one another by low ground, are thus formed. The Inferior Oolite gives rise to the range of hills that runs from Lamyatt Beacon, near Bruton, past Castle Cary and Cadbury Camp, to near Yeovil. A second ridge, formed by the Forest Marble, lies to the east and extends from Wanstrow, past Redlynch, to Bratton. South of this point it comprises the eminences of Charlton Hill and East Hill near Milborne Port and, outside the region, the ridge of Lillington Hill lying to the south-east of Sherborne. Between Yeovil and Crewkerne, the Forest Marble limestone forms the Abbott's Hill and Ashlands Hill ridge, which reaches a height of 182 m.

The drainage of the district is comparatively simple. Practically all rivers west of the Cotswold scarp flow into the Severn Estuary, whereas those which follow the dip slope of the Cotswolds join the Thames (Figure 2). The Bristol Avon is an ex-

Plate 1   Cheddar Gorge (A9756).

Plate 2 Dry valleys floored by permeable limestones.
a. Jurassic limestones in the Cotswold Hills (A10920).
b. Carboniferous Limestone, Burrington Combe, Mendip Hills (A10748).

Plate 3   The Wye valley, looking northwards from Symond's Yat to the Old Red
Sandstone ridge of Coppet Hill (A6262).

ception. Rising within the eastern margin of the district at Badminton, it flows
eastwards as a consequent stream down the dip slope of the Cotswolds for about
ten miles to Malmesbury, as if belonging to the Thames drainage system. At
Malmesbury it turns south-west and follows the strike of the Jurassic beds to
Bradford-on-Avon, whence it flows west and north to Bath, re-entering the district
at Limpley Stoke. In its passage to Bath it runs against the dip of the Jurassic rocks
and cuts a wide, deep valley. From Bath to Hanham Mills the river flows mainly
over Triassic and Lower Jurassic strata, and then cuts a gorge in Coal Measures
sandstone from which it emerges at Bristol. Here it turns as if heading for the Low
Flax Bourton Valley and the sea at Clevedon, but instead of following this apparently easy route along the outcrop of the soft Triassic rocks it turns north-west
at Clifton and enters the famous gorge carved in the hard Carboniferous
Limestone (Plate 4). Before flowing into the Severn Estuary at Avonmouth it is
joined at Sea Mills by the River Trym, which cuts a similar, though smaller gorge
through the Carboniferous Limestone ridge near Henbury.

The Avon and its tributaries thus appear to represent superimposed drainage,
i.e. their present anomalous courses were initiated not on the rocks and land surface at present exposed, but on a comparatively regular covering of Mesozic rocks
which was subsequently removed by denudation. The course of the River Wye
may have been similarly determined during the initial stages of its development.

SCENERY

Tremadocian and Silurian rocks form no special topographical features owing to
the limited extent of their outcrop and to the Triassic planation. By contrast, the

Plate 4   The Avon Gorge (A9763).

massive Brownstones and conglomerates of the Old Red Sandstone in Gwent, west of the River Wye, give rise to fine wooded scarps that sweep north-eastwards from Newport through Wentwood to near Trellech. Their soil is poor and thin, and the region is one of small farms with fields bounded by thick stone walls; much of the ground is now devoted to forestry. In the Forest of Dean, the conglomerates form several prominent ridges, such as Edge Hill, Soudley, and Coppet Hill near Symond's Yat (Plate 3). Old Red Sandstone rocks give rise to much of the wooded Failand ridge near Bristol, the sandy tract of Milbury Heath and the featureless moorland summit of Blackdown in the Mendips.

It is in the Carboniferous Limestone districts that some of the most striking scenery is to be found. Although the limestone is hard, it is traversed by strong vertical joints which determine the cliff profiles of deep gorges such as Cheddar Gorge (Plate 1) and the Wye Valley (Plate 5). These are incised in bare uplands, like the Mendips, where weathered limestone crags project through the thin soil. Owing to the solubility of limestone in waters charged with carbon dioxide, such tracts are usually waterless; streams reaching the limestone plunge underground through 'swallets' or 'slockers' and, running through subterranean passages and caverns glistening with stalactites, finally emerge in great springs at the foot of the hills. The underground streams and their springs are exemplified by the waters which pour into Eastwater Cavern, to emerge at Wookey Hole, and by those which descend the swallets at Charterhouse and issue at Cheddar.

The same phenomena may be seen, on a very much smaller scale, in some of the oolitic limestone disticts, where valleys are dry or are subject to seasonal flow over part of their course (Plate 2); but here the softness of the limestone precludes the formation of deep gorges.

Plate 5   The Horse-shoe Bend of the Wye from the Wyndcliff (A6278).

The scenery of the Coal Measures contrasts strongly with that of the Carboniferous Limestone. Consisting mainly of shale with a thick median division of sandstone know as the Pennant Formation, they give rise, in the Forest of Dean, to thickly wooded undulating ground of moderate relief in which the sandstone forms ridges. The scenery of the Bristol Coalfield is monotonous and, because of the late Triassic planation, it shows little of that differentiation of relief that might have been expected from the occurrence of alternating hard and soft rocks. In the Somerset Coalfield, however, the pre-Triassic surface has been uncovered in a few places only, and over the remainder of the area the Coal Measures are concealed by Mesozoic strata, which impress their individuality upon the scenery.

The predominantly soft nature of the Triassic and Liassic rocks gives rise to areas of low relief, such as the vales of Gloucester and Somerset. The scenery is not, however, without charm; the level country through which the River Severn flows forms miles of rich pasture and orchards, broken occasionally by undulating ridges of harder rock which at Tewkesbury, Westbury and Fretherne make low cliffs. In Somerset the Blue Lias limestones form the Polden Hills and other low ranges of hills separating the fertile plains of red Triassic marl from the heavy clay lands of the Lower Lias.

Southwards from Stroud, the Upper Lias sands crop out in the face of the Cotswold escarpment, and in south and east Somerset give rise to steep sandy slopes and knolls such as those of Glastonbury Tor and Montacute. Deeply sunken lanes or 'holloways' are a characteristic feature of this type of country.

A striking natural feature of the region is the prominent scarp of the Cotswolds, extending from Chipping Campden to Bath, formed by the Inferior and Great Oolite limestones. The former, dominating the escarpment in the north and mid

Cotswolds, is gradually replaced in importance in the south Cotswolds by the Great Oolite which forms the uplands overlooking the Avon Valley at Bath.

On the dip slopes, the limestones, falling gently eastward towards the Oxford Clay vales, give rise to a high undulating plateau, partly arable, partly downland, drained by the headwaters of the Thames and Avon, and providing shelter in its deep valleys for villages and mansions built of mellow freestone.

South of the Mendips no great single escarpment is present. Its place is taken by two step-like ridges; the lower and more westerly, formed by the Inferior Oolite, is separated from the higher scarp of the Forest Marble by a tract of Fuller's Earth Clay and Frome Clay in which the Fuller's Earth Rock makes a minor feature. East of the Forest Marble ridge lies a wide expanse of the heavy Oxford Clay, bordered on its eastern margin by the escarpments of the Upper Greensand and Chalk.

A widely different type of scenery, covering some 700 km^2, is that produced by the estuarine alluvium that flanks both sides of the Bristol Channel. The most extensive area lies south of the Mendips and is known as the Somerset Levels. Various parts of the area are known as Moors or Heaths, as for example Sedgemoor and Meare Heath. Much of the ground is below high-tide level and is protected from marine inundation by dunes of blown sand and sea walls. The rivers Axe, Brue, Parret and Cary (the last continuing seaward as the artificial King's Sedgemoor Drain) flow sluggishly westwards through an extensive region of pasture fields bounded by willow-lined ditches or 'rhines'. The flat landscape throws into relief the steep southern face of the Mendips (Plate 12), the inliers of Mesozoic rocks which form the Polden Hills and the striking residual hills of Glastonbury Tor and Brent Knoll.

EARLY GEOLOGICAL WORKS

Of all the names connected with the geology of the district none is more illustrious than that of William Smith, the 'Father of English Geology'. Whilst engaged in surveying the Somerset Coal Canal in 1792–95, he discovered the fundamental principles of stratigraphy, i.e. the constancy of the order of succession of the strata and the characterisation of each stratum by certain fossil species. In 1799 he coloured geologically the 'Map of Five Miles around the City of Bath', one of the oldest geological maps in existence. At Bath in June of the same year, he dictated his 'Tabular View of the Order of Strata' to his friends Townsend and Richardson.

The official Geological Survey maps at 1 inch to 1 mile of the Bristol–Gloucester district all appeared between 1845 and 1857 and were the results of the labours of such well-known men as Sir Henry T De la Beche, John Phillips, A C Ramsay, H W Bristow, W T Aveline, D H Williams and Edward Hull.

Mention must also be made of the remarkable set of 19 sheets of maps of the Bristol area, on the scale of 4 inches to 1 mile, produced in 1862 by William Sanders. These maps formed the basis of the local work of the 1871 Coal Commission in which John Anstie played a great part.

In addition to the official maps and publications, there are descriptions of the geology of the district by many private workers. The name of Weaver will be remembered for his observations on the Silurian rocks of Tortworth, and that of Charles Moore for his labours on the Rhaetic and Liassic rocks.

Much of our present knowledge of the Somerset Coalfield is due to the detailed work of James McMurtrie, whilst in literature on the Carboniferous Limestone the name of Arthur Vaughan stands pre-eminent. The lucid and stimulating papers of Charles Lloyd Morgan did much to inspire research into the geological history and structure of the Bristol district and had the effect of arousing great popular interest in the science.

Many well-known names figure in works upon the 'Oolites': Wright, Lycett, Hudleston and Witchell to mention a few, but the most famous is that of S S Buckman, whose work is referred to later.

2   Cambrian

The oldest rocks that crop out within the district belong to the Tremadoc Series, traditionally regarded in Britain as the youngest part of the Cambrian. These Tremadoc rocks occupy an inlier that extends northwards from the Tortworth area nearly as far as the River Severn at Tites Point (Figure 3). They are very poorly exposed and knowledge of the succession has mainly been gleaned from temporary exposures. They were first described by Smith and Stubblefield in 1933 and subsequently by Curtis (1968), while more recent mapping by BGS has thrown additional light on their structure. The succession is predominantly of micaceous shale, grey when fresh, with a variable proportion of interbedded siltstone or very fine-grained sandstone layers, lenticles and beds. The total thickness of the exposed rocks cannot be established with any accuracy. Calculations based on average dips and width of outcrop give a thickness of some 2200 m. This figure takes no account of tectonic thickening by folding and faulting, but a comparable thickness of mainly Tremadocian shale overlying Upper Cambrian shales and Lower Cambrian sandstone or quartzite has been proved in the Cooles Farm Borehole at Minety, less than 40 km to the east. Seismic reflection evidence suggests the extension of these rocks at depth along the entire eastern margin of the district.

The succession at Tortworth, which in broad terms youngs southwards, is divided lithologically into the Breadstone Shales below and the Micklewood Beds above. The base of the Tremadocian is not seen but the top is overlain unconformably, with slight angular discordance, by Silurian rocks; the Ordovician is not represented. Marine faunas have been collected from both formations and these show similarities with the North Wales Tremadoc faunas rather than with those of the geographically nearer Welsh Borderland and Midlands areas. It has been suggested, on the basis of numerous minor sedimentary structures in the sandy units, which are particularly abundant in the Micklewood Beds, that the rocks were deposited in relatively shallow agitated water.

Breadstone Shales

These comprise about two-thirds of the inlier. The interbedded siltstone layers are rarely more than a few centimetres in thickness. Over much of the area the fauna indicates a lower Tremadoc age (*Rhabdinopora* [*Dictyonema*] *flabelliformis* Zone) and includes horny brachiopods, bellerophontoid gastropods, trilobites (*Beltella, Niobella*) and ostracod-like forms, as well as the zonal index fossil. A record of the graptolite *Clonograptus sp.* suggests that the overlying *Clonograptus tennellus* Zone is represented in the northern part of the outcrop.

Micklewood Beds

These occur over the southern one-third of the inlier and, although the boundary with the Breadstone Shales is not clearly defined, are usually distinguished from

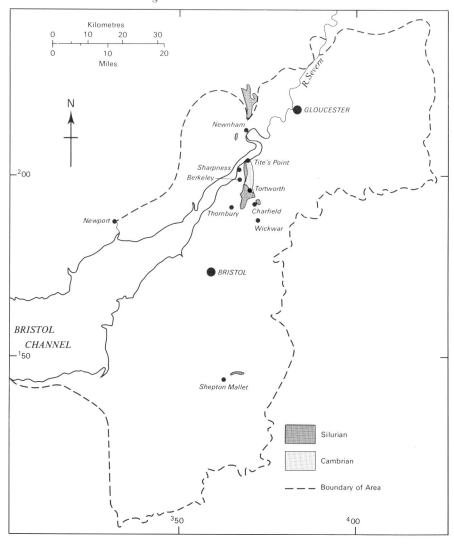

Figure 3   Outcrop of the Lower Palaeozoic rocks in the region.

them by the more common occurence of flaggy sandstone or siltstone beds. These beds are variable in thickness, but typically measure between 15 and 30 cm. The zonal age of the lowest 60 to 90 m is unknown, but the remainder of the succession is Upper Tremadoc in age. Some layers are crowded with fragments of the horny brachiopods *Lingulella* and *Schmidtites*, whilst the trilobites *Angelina* and *Peltocare* have also been recorded.

3 Silurian

Rocks of Llandovery to Ludlow age occupy a number of small inliers extending from north of the Severn near Newnham southwards to the eastern Mendips (Figure 3). The most extensive, and best known, is on the northern margin of the Bristol Coalfield in the Tortworth area. Although rocks of Silurian age have so far been proved only at depth under the Cotswolds, they are presumed to be present under much of the remaining part of the district. The rocks fall into four main divisions:

Přídolí (formerly Downton) Series (part)
Ludlow Series
Wenlock Series
Llandovery Series

The lower three series form a natural grouping of marine strata which, until relatively recently, comprised the Silurian system as understood in Britain. However, international agreement in 1972 to define the base of the overlying Devonian System at an horizon coincident with the base of the *Monograptus uniformis* graptolite biozone has meant that part or all of the overlying Přídolí Series must now also be included in the Silurian. No graptolites are, however, known from the Přídolí of Britain and no consensus of opinion has yet emerged as to where the top of the Silurian should be placed within the series.

The Llandovery – Ludlow rocks of this district comprise shallow-water, arenaceous and argillaceous, marine sedimentary rocks with some limestone. They accumulated in the southern part of a wide, intermittently, but gently subsiding shelf region that separated the rapidly subsiding Welsh Basin to the northwest from a land area called the Midland Block that lay to the east and the south. An important volcanic episode occurred in Somerset during the Wenlock.

The post-Ludlow strata of Silurian age fall within the lowest part of the Old Red Sandstone. Classically, in this country, the Old Red Sandstone has been accorded system status synonymous with the Devonian. In this district the Old Red Sandstone forms a natural grouping of continental facies sedimentary rocks, which, because it is largely Devonian in age, will be treated as a whole in the next chapter.

Widespread marine transgressions occurred in mid and late Llandovery times in Britain. The Llandovery Series of the type area at Llandovery has been divided, on the basis of brachiopod assemblages and lineages, into substages of which six, informally symbolised as C_1 to C_6, refer to the late Llandovery. The course of the marine transgressions across the shelf areas may be followed by the aid of these substages. In the region only C_5 and C_6 are present, represented by the Damery Beds and Tortworth Beds respectively. The strata, with a thin basal conglomerate, rest with slight angular discordance on the Tremadoc Series with no trace of Ordovician strata intervening. Relative, but not absolute water depths of deposition in the Llandovery and Wenlock, and hence bathymetric conditions, have been inferred by the study of marine, mainly brachiopod shelly communities in the Welsh Borderland areas (e.g. Hurst et al., 1978). By this and sedimentological criteria, it has been possible to deduce in general terms that the shelf seas deepened through late Llandovery times in the region, and that the Wenlock was heralded by marked shallowing followed by more variable conditions, leading to marked deepening again in early Ludlow times. Shallowing through the Ludlow led eventually to the establishment of continental conditions in the Přídolí.

TORTWORTH

The most complete succession of Silurian rocks has been established in the Tortworth and immediately adjacent Charfield inliers (Figure 4), though exposures are very limited. These rocks have long been recognised as Silurian, and Curtis (1972) has given an account of the history of research; modern knowledge mainly stems from his work and that of Cave (1977). The general succession is as follows:

Chronostratigraphy	Lithostratigraphy	Thickness (m)
Wenlock	Brinkmarsh Beds	
Series	Mudstone and sandstone	up to 30
	Upper Limestone	0 – 11
	Mudstone	c.120
	Middle Limestone	0 – 15
	Mudstone and sandstone	c.100
	Lower Limestone	0 – 30
Llandovery	Tortworth Beds	
Series	Mudstone and sandstone	c.105 – 300
	Upper Trap: basalt	0 – 75
	Damery Beds	
	Mudstone and sandstone	c.125 – 185
	Lower Trap: basalt	0 – 34
	unconformity	

The Llandovery sedimentary strata thicken westwards and the thickest sequence, about 500 m, is present in the north-western part of the inlier around Stone. Conversely, the igneous strata are absent in the west but thicken southeastwards (Figure 4).

The Lower Trap and Upper Trap

These volcanic rocks were originally thought to be intrusions, but the balance of evidence now favours an extrusive origin. They are both contaminated enstatite-bearing basalts, with or without olivine. Both may be strongly amygdaloidal and are much altered, with plagioclase completely or partially albitised, enstatite replaced by bastite and olivine altered to chlorite and opaque oxides. In the Upper Trap the presence of quartz xenocrysts, which are clearly visible with a hand lens, usually provides a ready means of distinguishing the two flows in the field.

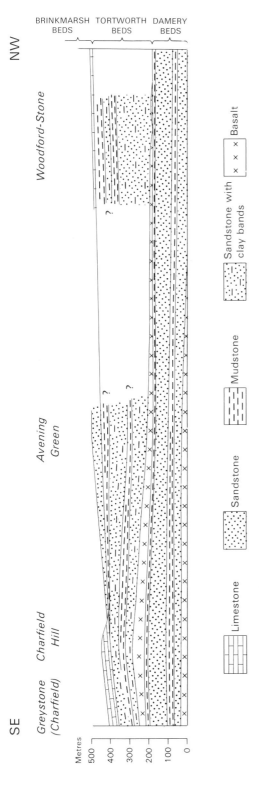

Figure 4   Interpretative diagram of thickness changes in the Llandovery along its outcrop between Charfield and Stone (after Cave, 1977, fig. 5).

Damery Beds

The Damery Beds consist of weakly calcareous, thin-bedded, sandstone, siltstone and mudstone, with occasional layers of impure limestone, which are usually associated with fossil-rich bands. The colours range from grey to green, with ?Triassic staining giving rise to reds and purples. For mapping purposes it is usually possible to subdivide the strata into thick members, each dominated by either sandstone or mudstone. Fossils are abundant throughout, with brachiopods, often in great numbers, being by far the commonest forms. Crinoid columnals, *Tentaculites,* corals and trilobites are plentiful; gastropods and bivalves less so. No faunal subdivisions of the Damery Beds have been established, but there is a well-marked change in the composition of the fauna towards the top, thought to be related to increasing water depth. In the lower part of the Damery Beds the brachiopod *Eocoelia curtisi* is abundant, accompanied by small rhynchonellids. These forms continue into the middle and upper parts of the formation but become less conspicuous. Other forms, like *Leptaena rhomboidalis, Costistricklandia lirata alpha, Atrypa reticularis* and *Howellella anglica* become more numerous. Some species such as *Leptostrophia compressa* and the trilobite *Dalmanites weaveri* are fairly common throughout the succession.

Tortworth Beds

The Tortworth Beds are less well known than the underlying strata They appear to be of generally similar lithology, but tend to be less fossiliferous. Commonly present at the base is the Palaeocyclus Band, a thin decalcified sandstone characterised by the button coral *Palaeocyclus porpita,* the tabulate coral *Favosites,* abundant brachiopods and *Dalmanites weaveri.* Where the Upper Trap is present there is commonly a metre of ashy limestone with a large fauna, including abundant tabulate corals intervening between the top of the lava and the Palaeocyclus Band. The fossils from the lower horizons indicate the *Costistrictlandia* community (that is a group of animals living in the same habitat), while higher levels probably contain representatives of the shallower water *Eocoelia* community.

Brinkmarsh Beds

The Brinkmarsh Beds, of Wenlock age, consist predominantly of grey mudstone with thin layers and beds of siltstone and fine-grained sandstone. The sandstone is calcareous and the mudstone may occasionally include layers of calcareous nodules. The sandy beds may show ripple marks, cross-bedding, drag marks and 'ball-like' load structures. The succession includes three units of argillaceous limestone which are usually crinoidal and very fossiliferous. They are thin-bedded, rubbly in texture and grey in colour, except where stained red due to proximity of the Triassic rocks. The limestone units give rise to ridges in the otherwise somewhat subdued topography determined by the intervening mudstone.

The base of the Brinkmarsh Beds is taken at the base of the lowest limestone unit or, where this is absent, beneath a series of red-stained calcareous sandstone beds containing abundant crinoid columnals, apparently a local facies variation of the limestone. The basal limestone unit and its associated sandstone appear to be practically continuous throughout the inlier, in contrast to both the middle and upper limestone units, particularly the latter, which tend to be lenticular in development. The lowest limestone unit is clearly equivalent, in part at least, to the well-known Woolhope Limestone of the areas to the north, but the others appear to be local developments not represented elsewhere. The highest exposed Wenlock strata are

in the Milbury Heath area, at the southern end of the inlier, where they are unconformably overlain by rocks of late Devonian age. They contain a fauna that is common to both the Wenlock and Ludlow elsewhere.

The total thickness of exposed Brinkmarsh Beds approaches 300 m, but due to
the late Devonian unconformity the full succession is not present at outcrop. It is
probable that the unexposed strata at the top of the succession lie beneath rocks of
Ludlow age to the north-west of the inlier and contain a representative of the Wenlock Limestone, so widespread to the north.

SHARPNESS

A narrow, faulted outcrop of rocks of Ludlow age extends inland from Tites Point
for some 5 km on the downthrow (western) side of the north–south-trending
Berkeley Fault to within 2 km of the Silurian outcrops of the Tortworth inlier. On
the foreshore at Tites Point, 22 m of 'Whitcliffe Beds' overlying perhaps a further
11 m of 'Leintwardine Beds' are exposed in a faulted, plunging anticline. These
rocks, of Ludfordian age, are overlain by strata attributed to the Downton (now
Přídolí) Series (Cave and White, 1971) on both sides of the anticline.

About 2 km inland, the Brookend Borehole proved a nearly complete Ludlow
succession beneath the Přídolí. The formational names used in it are derived from
the type Ludlow area, with which correlation was made mainly by means of the
abundant shelly faunas, dominated by brachiopods. The dip of bedding in the
borehole varies from an average of 56° above 177 m depth to 35° below, and the
apparent thicknesses given in the borehole log have been corrected to give true
thicknesses, calculated to the nearest metre and shown in brackets after the
borehole thicknesses. A total of about 160 m of Ludlow strata was proved. The succession is as follows:

		Thickness m	Depth m
Přídolí Series	THORNBURY BEDS Mudstone, dark brownish red, silty, with small dolomitic concretions at several levels; thin sandstone layers with *Lingula*, fish and plant remains	83.29 (c.47)	83.29
	DOWNTON CASTLE SANDSTONE Sandstone, mainly grey, cross-bedded, slightly glauconitic, with bone beds; scattered marine fossils including *Lingula* and fish fragments	7.34 (c.4)	90.63
	LUDLOW BONE BED Siltstone and fine sandstone with phosphate pebbles and fish fragments	0.15 (0.08)	90.78
	'WHITCLIFFE BEDS' Mudstone, grey, with siltstone layers and sporadic thin fragmented and shelly limestone layers; 5 cm basal conglomerate	21.44 (c.12)	112.22

	'LEINTWARDINE BEDS' Mudstone, shelly with wisps and layers of calcareous siltstone, and thin beds of shelly limestone and conglomeratic limestone, particularly below 158.5 m	55.88 (c.32)	168.10
Ludlow Series	'BRINGEWOOD BEDS' Mudstone, grey, with common fossiliferous, argillaceous, nodular limestone layers; corals above 206 m	60.50 (c.50)	228.60
	'ELTON BEDS' Mudstone, grey, uniform, poorly fossiliferous, with rare limestone layers	86.31 (c.70)	314.91

The lowest strata contain a fossil assemblage indicative of deep-shelf conditions and thought to be characteristic of the lowest part of the Ludlow Series. The succeeding rocks reflect progressive shallowing of the shelf sea, with depositional pauses marked in the 'Leintwardine Beds' by conglomerates with bored limestone pebbles, believed to represent hard-grounds. Sporadic layers of bentonite throughout the successions appear to indicate volcanic activity, though the sediment source is not known at present. The Downton Castle Sandstone testifies to the final shallowing of the sea before the onset of dominantly continental conditions represented by the overlying Thornbury Beds.

NEWNHAM

A small inlier of Ludlow rocks is situated about a kilometre south-west of Newnham, north of the River Severn. The beds, seen in two old quarries, are inverted and dip to the east between 59° and 75°; Cave and White (1971) found that the best section was in the southernmost quarry. This showed some 20 m of strata consisting of 11 m of 'Whitcliffe Beds' and the uppermost 9 m of the 'Leintwardine Beds'. The rocks are similar to those already described in the Sharpness inlier to the south.

WICKWAR

West of Wickwar, in the valley of the Little Avon, there are small exposures of Wenlock strata showing dips of around 30° in a general south-westerly direction. The rocks consist of pink dolomitic siltstone and argillaceous limestone. An adjacent shaft proved about 8 m of very fossiliferous limestone.

MENDIPS

To the north-east of Shepton Mallet, Silurian rocks form a narrow belt in the core of Beacon Hill Pericline and are overlain with low angular unconformity by late Devonian (Old Red Sandstone) strata. In the middle of the inlier the rocks are exposed in several large quarries where the strata are vertical or very steeply dipping and young to the north. The area has recently been described in detail by Hancock (1982), who disagreed with the interpretation of the structure as an anticline and gave a complete succession across the inlier as:

		Thickness
		m
11	Andesite lava	5
10	Agglomerate	18
9	Andesite lavas	70
8	Tuff and bedded agglomerate	20 – 29
7	Andesite lavas	90 – 135
6	Tuff with red and green mudstone	18
5	Andesite lavas	50
4	Tuffs, locally fossiliferous, sandy tuffs and red mudstones	105 – 135
3	Andesite lavas	30
2	Tuffs	34 – 60
1	Fossiliferous Wenlock shales	95

The total thickness is about 600 m. It has now been shown that the fossiliferous Wenlock shales are the oldest part of the succession, not the youngest as had hitherto been thought. They have been dated by the brachiopod *Eocoelia angelini* as being of early Wenlock age. Shallowing of the sea during the succeeding volcanism is suggested by the shelly assemblages present in Bed 4, and there is evidence for both subaerial and submarine eruption in different parts of the succession. The lavas are massive flows of highly altered pyroxene-andesite. Lithologically, the pyroclastic rocks range from fine-grained, well-bedded, water-lain tuffs to massive, coarse agglomeratic beds with boulders and bombs of lava up to 30 cm across; colours range from blue-black to green and red, but with shades of buff and brown predominating. A large agglomerate mass that cuts all the members of the succession to the east of the main quarries is interpreted as a vent-fill.

BATSFORD (LOWER LEMINGTON) BOREHOLE

The extent of the Silurian rocks beneath the Upper Palaeozoic and Mesozoic cover east of the inliers is not known but they were found in the Batsford Borehole, situated 2 km south-east of Moreton-in-Marsh (Williams and Whittaker, 1974). This borehole proved more than 52 m of limestone, shale and sandstone, with dips of 10° to 49°, beneath gently dipping Coal Measures (see also pp.63, 65). The shelly fauna (Substage C_6) indicates a late Llandovery age.

4   Old Red Sandstone (Silurian – Carboniferous)

Rocks of the continental red-bed facies known as the Old Red Sandstone have their most extensive and complete development in the part of the region west of the River Severn (Figure 5). Here, the rocks are exposed around the Forest of Dean coalfield in major, north – south-trending asymmetric folds. East and south of the River Severn the Old Red Sandstone is restricted in occurrence to the partly faulted Sharpness – Thornbury inlier, a narrow outcrop around the periphery of the Coalpit Heath syncline, and to the cores of anticlines around Bristol and in the Mendips.

Table 2 shows the main features of the Old Red Sandstone stratigraphy of the district. The position of the Silurian/Devonian boundary within the Old Red Sandstone is unknown, although a thin, but widespread, composite air-fall tuff sequence, recently recognised in the upper part of the Raglan Marl (or its equivalents), may provide an acceptable horizon for regional correlation and be useful as a reference point for the boundary. The standard Devonian stages relate to a marine succession and are defined in terms of European ammonoid and graptolite zones, supplemented by conodont faunas. The continental Old Red Sandstone, on the other hand, has classically been related to series defined in terms of zones based on fish faunas. Knowledge of the palynology of the red beds is not, as yet, sufficient to provide firm stratigraphical conclusions applicable to the district. The correlation between the marine and continental chronostratigraphical schemes is only approximate. Furthermore, in the region even correlations within the Old Red Sandstone must be treated with caution due to the sporadic and often only local development of the vertebrate faunas.

Sedimentation of continental facies deposits continued without significant break from the Silurian to about the end of early Devonian times (i.e. the end of the Emsian age or near the end of the Breconian epoch) in an area known as the Anglo-Welsh Platform, which approximated to the area formerly occupied by the Silurian shelf sea. Open sea at this time lay at about the southern limits of the region. Although evidence from current directions throughout the lower Old Red Sandstone suggests derivation of the thick terrigenous sedimentary succession from an uplifted, eroding area to the north and north-west (known as St George's Land), the first evidence within the region of this comes from the pebble content of conglomerates of mid-Breconian age near the top of the lower Old Red Sandstone.

Earth movements, uplift and concomitant erosion followed during middle Devonian times. Deposition of continental red-beds resumed across the district in late Devonian times. The southern seas encroached northwards at this time, but although the earliest marine incursion into the south Midlands may have extended into the eastern part of the region (see p.64), it was not until Carboniferous times that the region was once more wholly under the sea. The contact between the lower and upper Old Red Sandstone strata takes the form of a disconformity and, even

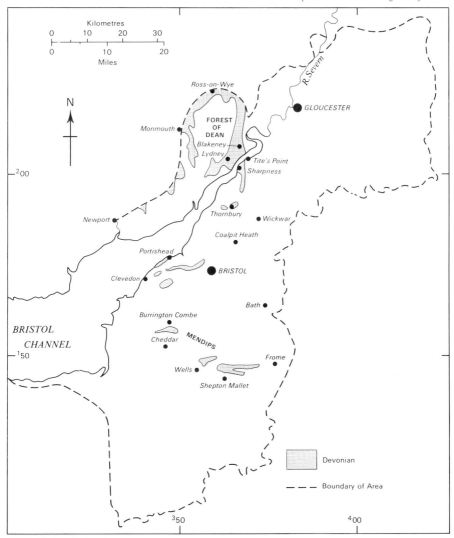

Figure 5   Outcrop of the Devonian rocks in the region.

where great thicknesses of sedimentary rock can be proved to have been removed in the interval, the junction is typically nonangular.

Jawless, armoured ostracoderms are dominant in the Lower Old Red Sandstone fish faunas. In ascending order, zones are based on species of *Hemicyclaspis, Traquairaspis, Pteraspis* and finally, at the top of the Dittonian, *Althaspis*. Rocks of Breconian age contain only rare vertebrates. By late Devonian times these primitive faunas had been completely replaced by more advanced ganoid fish of which *Holoptychius* and *Bothriolepis* are the best known. Plant remains occur throughout the Old Red Sandstone, but are of too limited occurrence to be of critical stratigraphical value.

The sedimentology, including conditions of deposition, has been studied in much detail in this and adjacent regions over the last twenty years (for references

Table 2   Classification and correlation of the Old Red Sandstone of the Bristol–Gloucester region, partly after Allen *in* House et al. (1977) (not to scale).

System/Subsystem	Stage	West of Severn	East of Severn — BRISTOL AREA / TITES POINT AREA		Old Red Stages	
Upper Devonian — C	Fammen-nian	Tintern Sandstone (60–150 m) Quartz Conglomerate (6–30 m)	⎡Tintern Sandstone (90–120 m)⎤ north of ⎣Quartz Conglomerate (c.10–15 m)⎦ Bristol *passing southwards into* Portishead Beds (220–500 m) Bristol & Mendips		Farlovian	Upper Old Red Sst.
Upper Devonian	Frasnian					
Middle	Givetian					
Middle	Eifelian to Cou-vinian					
Lower Devonian	Emsian	Brown-stones (0–1100 m)	Black Nore Sandstone (500 + m)		Breconian	Lower Old Red Sandstone
Lower Devonian	Siegen-nian	St Maughans Formation (380–c.600 m)	St Maughans Formation (200 + m)		Dittonian	
Lower Devonian	Gedin-nian	Raglan Mudstone (385–c.600 m)	[*not seen*] Thornbury Beds (0–600 m)			
Silurian (part)		Downton Castle Sandstone (15–25 m)	[*not seen*] Downton Castle Sandstone (0–4 m)		Downtonian	
Silurian (part)		[Ludlow]	[*not seen*] [Ludlow to Wenlock]			

C = Carboniferous

see Allen and Williams, 1979). Cuttings along the M50 motorway (immediately north of the district) have provided relatively continuous exposure, hitherto lacking, through the greater part of the Lower Old Red Sandstone (Allen and Dineley, 1976). In broad terms, the Lower Old Red Sandstone consists of a progressively upward-coarsening sequence of clastic sedimentary rocks with marine influence in the lowest part (p.23), whereas the Upper Old Red Sandstone is an upward-fining sequence exhibiting a marine influence at the top.

STRATIGRAPHY

LOWER OLD RED SANDSTONE

Downton Castle Sandstone

The base of the Old Red Sandstone is marked by the well-known Ludlow Bone Bed, a thin, littoral lag deposit at the base of the Downton Castle Sandstone; it marks a probably small hiatus following deposition of the underlying Ludlow rocks. The type area of the sandstone formation is around Ludlow, but comparable, though not necessarily completely synchronous sandstones are present throughout the Welsh Borderland and South Wales. They were laid down under marine littoral conditions in either a delta or coastal plain during the final stage in the shallowing of the Silurian shelf seas. These rocks contain lingering elements of the shelly Ludlow fauna, including thin beds with *Lingula* and, more rarely, ostracods, bivalves and gastropods, with an admixture of the fish faunas that characterise the overlying strata. The rocks have been termed 'Grey Downtonian' in contrast to the 'Red Downtonian', applied to the overlying Raglan Mudstone Formation.

Raglan Mudstone Formation and St Maughans Formation

The Downton Castle Sandstone is sharply succeeded by the Raglan Mudstone Formation west of the River Severn, and the Thornbury Beds east of the Severn. Unlike some areas to the north, the basal junction of these two formations is sharp. They comprise thick, blocky, faintly laminated, red-brown silty mudstone interbedded with subordinate green, brown and purple sandstone. The sandstone commonly includes mudstone clasts and may infill mudcracks in the underlying mudstone. Deposits of air-fall tuff occur in the upper part of the Raglan Mudstone. The overlying St Maughans Formation is generally similar, but the proportion of sandstone to mudstone, approximately 1:3, is three or four times as high as in the underlying formation.

In these formations, the rocks are typically arranged in upward-fining sequences of intraformational conglomerate, channel sandstone and, finally, mudstone. The depositional environment was a wide alluvial plain intersected by rivers of moderate to high sinuosity draining towards the south. The positions of the river and stream channels are marked by the arenaceous rocks. At the bases of the thicker channel deposits there are laterally extensive pebbly beds with well-marked erosive bases. The pebbles are of local rock types, including abundant red marl. Disarticulated fish debris may be locally abundant.

Perhaps the most distinctive sedimentological feature of the Lower Old Red Sandstone is the presence, mainly within mudstone, of beds containing abundant micritic, calcitic or dolomitic concretions. These take the form of rounded nodules and subvertical, branching pencil-like forms. Aggregation of the nodules led to the formation of rubbly limestone beds known as 'cornstone'. There is every gradation between vaguely defined beds or zones of mudstone with concretions to limestone beds. These may form cappings to concretion-rich zones ranging from a metre to 5 m or more in thickness. After decades of argument it is now recognised that these carbonate sediments correspond to present-day caliche or calcrete deposits which are precipitated during soil-forming (pedogenic) processes under hot arid conditions with low seasonal rainfall. They testify to long periods when extensive surfaces of the exposed interfluves of the alluvial plains were stable and starved of sediment. Flash flooding and surface sludging, following sporadic, locally intense storms caused the breaking up of the surface layers to give rise to cornstone-breccias and conglomerates which are distinct from the intraformational conglomerates that formed in the beds of less ephemeral streams (Allen and Williams, 1979).

Calcrete horizons occur most abundantly in the upper parts of the Raglan Mudstone and the lower parts of the St Maughans Formation and reach their maximum development in the widespread Psammosteus Limestone at the top of the former. A less well-developed sequence of carbonate-bearing beds also occurs at the top of the St Maughans Formation.

Brownstones

The St Maughans Formation passes up into the Brownstones west of the River Severn, and the Black Nore Sandstone on the east, without any sedimentary break, by upward diminution of the content of interbedded mudstone. In the lower part of the formation, mudstone accounts for about one-third of the total thickness, but higher up the proportion of mudstone becomes negligible and the sandstone becomes coarser grained. The sandstone is red and brown in colour, well sorted, fine to coarse grained, with several types of cross-bedding, parallel lamination and cross-lamination. The sandstone occurs in upward-fining cycles of from a few to many metres in thickness, commonly with conglomerate at the base and underlain by an extensive erosion surface. The sandstone was deposited in a low-sinuosity (braided) river system draining towards the south. The deposits represent the medial parts of the Old Red Sandstone drainage system, in contrast to the underlying formations which represent the distal parts. The conglomerates in the higher part of the formation may contain abundant far-travelled pebbles derived from the erosion of Palaeozoic and Precambrian terrains.

UPPER OLD RED SANDSTONE

The Upper Old Red Sandstone has not received the same detailed study as the older rocks. In the north of the region the succession is divided into the basal Quartz Conglomerate and the overlying Tintern Sandstone, composed predominantly of interbedded sandstone, conglomerate and mudstone. Around Bristol and the Mendips, the much thicker succession is designated the Portishead Beds. Throughout the region the uppermost part includes interdigitations of fossiliferous marine shales and limestones indicating a coastal depositional environment. The fossils show that these passage beds are Carboniferous in age. The provenance of the terrigenous sediment appears to be similar to that of the Lower Old Red Sandstone.

WEST OF RIVER SEVERN

The Downton Castle Sandstone does not outcrop in this part of the region because of faulting, though it is present immediately to the east and west in the May Hill and Usk inliers respectively. The Raglan Mudstone gives rise to gently undulating ground, commonly with topographical features formed by the cornstone and sandstone beds. The best sections within the district occur in the coastal area between Blakeney and Lydney. Strongly indurated tuffaceous mudstone, estimated to lie about 100 m below the top of the formation, has been mapped in the area between Raglan and Monmouth where it caps steep ridges of mudstone. Allen and Williams (1981) have correlated it with the Townsend Tuff of Pembrokeshire and other occurrences over a wide area from Pembroke to the Clee Hills. The deposit, which is about 3 m thick, comprises three separate but closely spaced, graded, air-fall tuff beds, each with distinctive characteristics. The rocks are porcellanous and vary from green to white and purple. The Psammosteus Limestone is best developed in the area south-west of Monmouth, where the capping calcrete may attain 3 m or more in thickness and gives rise to prominent scarp features. Subsidiary calcrete horizons occur above and below this one. The total thickness of Raglan Mudstone in the M50 section is about 385 m, but it thickens southwards and, adjacent to the River Severn, appears to be at least 600 m.

The St Maughans Formation gives rise to hillier and higher ground than the Raglan Mudstone. The rocks are, in general, not well exposed, though there are numerous old quarries in sandstone and cornstone. The M50 section proved a total thickness of 695 m of strata, but southwards the formation has thinned to between 380 m and about 500 m.

The Brownstones form a striking scarp in the western and central parts of the area. On the east side of the Forest of Dean they dip steeply westwards and form a series of prominent north – south ridges from Mitcheldean southwards past Blakeney to Lydney. Fossils are sparse, apart from trace fossils in the finer-grained facies, and the age of the beds is uncertain. In the adjacent area to the west of the district the lowermost beds have yielded Breconian plants and Siegenian spores, whereas beds higher in the sequence near Mitcheldean have yielded late Dittonian fish remains. This apparent contradiction remains to be solved satisfactorily. A study (Allen, 1974) of far-travelled pebbles from the upper part of the Brownstones at Ross-on-Wye has shown that a wide variety of rock types is represented. They include acidic lava and tuffs, some of the latter with Ordovician graptolites, greywacke and various sandstones, some preserving Silurian, shelf-facies shelly faunas and others of 'Grey Downtonian' aspect. Metamorphic rocks, including mylonite, jasper and quartzite, invite comparison with the Precambrian rocks of Anglesey. Allen concluded that the assemblage originated in the Welsh region and provided the first evidence of uplift, denudation and the partial removal of early Devonian and Silurian rocks in that region. The Brownstones have been much quarried and one of the best sections is in the Wilderness Quarry east of Mitcheldean, which has yielded an abundant fish fauna, including the late Dittonian ostracoderm zonal fossil *Althaspis leachi.* The formation is thickest in the north-eastern part of the Forest of Dean where it attains about 1100 m. Elsewhere it is diminished by intra-Devonian or intra-Carboniferous overstep.

The Quartz Conglomerate, at the base of the Upper Old Red Sandstone, commonly makes a well-marked, in places craggy feature. The formation, from 2 to 30 m thick, comprises one to several thick conglomerate beds interbedded with sandstone. Individual conglomerate beds may reach 6 m in thickness, such as at Symonds Yat where the formation is at its thickest. Pebbles are largely vein quartz with some quartzite, jasper and decomposed igneous rocks. Together with the overlying Tintern Sandstone, the Quartz Conglomerate forms high upland tracts of poor sandy soil, usually devoted to forestry. The Tintern Sandstone, up to 100 m thick, consists of interbedded grey-green, red and purplish brown sandstone with subordinate quartz-pebble conglomerate and red and green mudstone. At the top, passage beds of alternating sandstone, shale and limestone grade up into the Carboniferous Lower Limestone Shale. These passage beds are 14 m thick at Drybrook.

SHARPNESS – THORNBURY

The Downton Castle Sandstone is exposed on the foreshore at Tites Point beside the River Severn. The beds, which comprise 1.7 m of grey-brown to reddish brown, mostly soft, thin-bedded sandstone, rest non-sequentially on Ludlow mudstone and siltstone (see above). The sandstone fills a shallow incised channel in the Ludlow rocks, and there is a representative of the Ludlow Bone Bed in the centre of the channel. The sandstone is sharply overlain by the red-brown silty mudstone of the Thornbury Beds. The best section through this formation, however, is in the Brookend Borehole (p.17 – 18). The name Thornbury Beds has been coined because the Psammosteus Limestone, which marks the top of the Raglan Mudstone, has not been found on this side of the River Severn and the possibility cannot be ruled out that part of the St Maughans Formation is represented in the upper part of the Thornbury Beds. The latter are present in discontinuous sections for some 6 km down the coast from Tites Point. Inland the beds give rise to a subdued outcrop, marked by discontinuous low ridges formed by the sandstone beds. The only firm evidence of higher formations of the Lower Old Red Sandstone is in the Severnside Borehole, about 1 km south of Pilning, in which 450 m of strata, divided between the Brownstones and the St Maughans Formation, were proved.

The Upper Old Red Sandstone succession is similar to that to the west of the River Severn. The most complete section is in a road cutting at Buckover, northeast of Thornbury, at the southern end of the Tortworth Inlier. The discovery, here, of a fish-bed with *Bothriolepis*, only 1.2 m above the base of the Quartz Conglomerate, provides evidence that the Upper Old Red Sandstone lies entirely within the Farlovian Stage. Middle Devonian earth movements reach their maximum known expression north of Bristol in the Tortworth Inlier, where a combination of gentle folding, horst development and erosion resulted in the Quartz Conglomerate resting directly on early Wenlock strata in the middle of the inlier. Small exposures of Thornbury Beds occur at Wickwar on the eastern side of the horst.

BRISTOL

In the Bristol area, the Lower and Upper Old Red Sandstone occur in the core of the Westbury-on-Trym Anticline and in the Clevedon – Portishead ridge. The type area for the Black Nore Sandstone and the Portishead Beds is on the northwest side of the latter ridge, where the rocks are well exposed along the coast. Lithologically, the Black Nore Sandstone is comparable with the Brownstones of the main outcrop west of the River Severn; in the absence of fossils, however, the correlation cannot be proved.

The Upper Old Red Sandstone has not here been subdivided into two formations and the whole sequence of red and reddish purple sandstone with subordinate, interbedded red and green mudstone, siltstone and conglomerate is given the name Portishead Beds; all the beds are lenticular. The formation varies from about 240 to 275 m thick in the Bristol – Portishead area. Apart from the Avon

Gorge, the best exposures are seen in old stone quarries, where some of the beds were worked for building stone.

The lowest member of the Portishead Beds is a coarse, unsorted conglomerate, 3 to 5 m thick, known as the Woodhill Bay Conglomerate after its exposure in Woodhill Bay. The constituent pebbles are mostly augen-quartz and dark brown quartzite, but jasper, chert, quartz, mica schist and silicified igneous rocks have also been recorded, which suggests derivation from a Precambrian terrain (Wallis, 1927). The junction with the underlying Black Nore Sandstone is a sharp, irregular erosion surface (Plate 7A, see p.81). A calcrete extends into the Woodhill Bay conglomerate and downwards for about 1 to 2 m below the contact. Important fish faunas have been collected from the Woodhill Bay Fish Bed, a 10 m-thick, mainly siltstone unit about 32 m above the base of the Portishead beds at Woodhill Bay, and from a conglomerate known as the Sneyd Park Fish Bed at the top of the sequence in the Avon Gorge to Westbury-on-Trym area. The overlying Shirehampton Beds are passage beds transitional to the Lower Limestone Shale, which have been traditionally included with the Carboniferous Limestone (p.34). Palaeocurrent data indicate derivation mainly from the north-west in both the Black Nore Sandstone and the Portishead Beds.

BATH

The Hamswell Borehole, some 5 km north-west of Bath, proved 293 m of Thornbury Beds without reaching their base. The lowest beds contain a Downtonian fish fauna. As elsewhere, it is thought that the Psammosteus Limestone may be absent and that Dittonian strata could be represented in the higher part of this formation. This occurrence may occupy a horst analagous to that of the Thornbury Beds near Wickwar to the north (see p.26).

MENDIP HILLS

Old Red Sandstone strata occur in the cores of four periclines in the Mendips (p.68) and in small inliers south-east of Cheddar and north of Frome. The rocks are poorly exposed, but appear to belong entirely to the Portishead Beds, of similar facies to that of the Bristol – Portishead area. The base is seen only in the Beacon Hill pericline. Here, the beds overlie Wenlock strata with an angular unconformity of 10° or more. This area is due south of Tortworth where the upper Old Red Sandstone is also unconformable on Wenlock strata (p.17). The succession is about 410 m thick, in the lowest 50 m of which are prominent conglomeratic and pebbly beds. The main body of the formation consists of red sandstone, but in the top 75 m grey sandstone and thick beds of red and green mudstone are abundant. A similar succession, about 500 m thick without the base being proved, appears to be present in the Blackdown Pericline. A fish- and plant-bearing conglomerate has been recorded at Burrington Combe 25 m below the top of the formation, possibly correlated with the Sneyd Park Fish Bed of Bristol, whilst spores in the uppermost beds suggest an early Tournaisian (Carboniferous) age. Unlike the areas to the north and south, there are no passage beds and the junction with the overlying Lower Limestone Shale (Carboniferous) is sharp.

5   Lower Carboniferous (Dinantian)

The Dinantian rocks of this district have long been known as the Carboniferous Limestone Series. Used in a lithostratigraphical sense, Carboniferous Limestone remains acceptable as a collective name for this group of rocks.

The Carboniferous Limestone has the most extensive outcrop of any Palaeozoic rocks within the district and forms much of the high ground and the most striking scenery. In the north of the district, its outcrop (Figure 6) surrounds the Forest of Dean Coalfield, except in the south-east, where it is concealed by an overstep of the Coal Measures and extends south-westwards to Chepstow in the narrow Tiddenham Chase Syncline, and then westwards to Magor in the broad Caerwent Syncline. Carboniferous Limestone rims the northern part of the Bristol Coalfield, extending from Over north-eastwards to Tortworth and thence southwards to Chipping Sodbury. Along the strike south of here, it is concealed by newer rocks, except in small inliers near Codrington and Wick.

West of Bristol, Carboniferous Limestone forms the high ground, cut by the Avon Gorge, extending from Penpole Point through King's Weston to Durdham Downs and thence by Failand to Clevedon. It also gives rise to the Clevedon–Portishead ridge. Carboniferous Limestone constitutes the dome-like structure of Broadfield Down and, farther south, is the principal rock group in the Mendip Hills, with an outcrop extending 50 km from Frome westwards to Brean Down. Beyond there, it continues out to sea in the islands of Steep Holm and Flat Holm.

Except for small inliers protruding through the Mesozoic cover under the southern lee of the Mendips, and the isolated inlier of Cannington Park near Bridgwater, no Carboniferous beds are exposed south of the Mendips in the region.

CLASSIFICATION

The Geological Survey officers who carried out the original survey of the district, and also subsequent workers, notably Wethered, working in the Forest of Dean, and Lloyd Morgan, in the southern part of the Bristol Coalfield, classified the Carboniferous rocks on a lithostratigraphical basis. This method was abandoned following Vaughan's classic paper on the Avon Gorge section in 1905. Vaughan grouped the Lower Carboniferous rocks into the Avonian Series, which he subdivided into five zones based on what he considered were evolutionary lineages, especially of corals. The zones were symbolised by the letters K, Z, C, S and D. Application of this zonal scheme by Vaughan and many other workers was undertaken with enthusiasm over much of Britain in the succeeding decades, but in the process, both the underlying concept and the zones themselves underwent extensive alteration in an attempt to achieve consistency between different areas. When

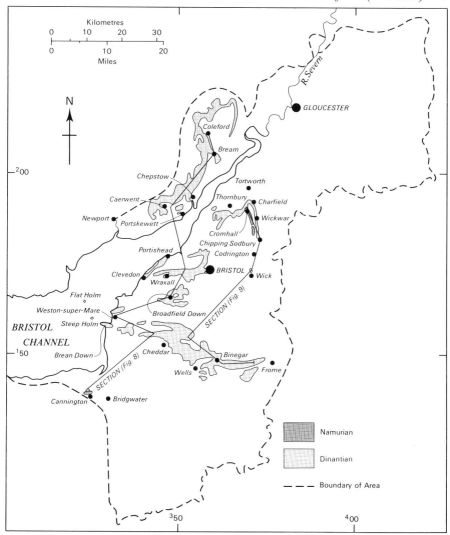

Figure 6 Outcrop of the Dinantian and Namurian rocks in the region.

the district was remapped by the Geological Survey, starting in the Forest of Dean in 1933, Vaughan's scheme was abandoned and replaced by a new, lithostratigraphical nomenclature (Kellaway and Welch, 1955).

Following pioneer work on the early Carboniferous rocks by Dixon and later workers, mainly in South Wales, the concept of 'bathymetric cycles' (i.e. cyclical alternations of shallower- and deeper-water sedimentation) evolved. This was developed by Ramsbottom in several papers between 1973 and 1976. He proposed a classification based on six major transgressive events, separated by periods of regression. These were recognised on the basis of lithology and fossil content. Ramsbottom's major cycles approximately coincide with stages proposed by the Dinantian Working Group of the Geological Society of London (George et al., 1976) because each major transgression was accompanied by the migratory faunas that are used to recognise the different stages. These stages have been defined in terms of type sections (stratotypes) and are applicable only in Britain. The application of the different classification schemes to the Avon section (Figure 7) shows the close relationship between conditions of deposition, lithology and faunal content of the strata.

SEDIMENTATION

The open sea, which during late Devonian times lay to the south and east of the district, spread over the whole area during earliest Carboniferous times. In the Mendips, the change from continental to marine conditions was abrupt, whereas to the north and south, passage beds, in which Old Red Sandstone facies rocks are interbedded with marine shale and limestone, indicate a period of fluctuating shorelines. After this relatively short initial period, however, the low-lying Old Red Sandstone alluvial plains were submerged by the sea, leaving a land mass, known as the St George's Land–Brabant massif, stretching from Ireland across Britain to Belgium, some distance to the north of the present region. It was in the tropical shelf seas bordering this massif that the Carboniferous Limestone was deposited. At first, much terrigenous material, represented by parts of the Lower Limestone Shale, was carried into the sea and deposited, but with the change to a high energy environment, deposition of fine clastic material ceased, the sea water cleared and carbonate sedimentation became dominant, both by the accumulation of the hard parts of the teeming marine life and by chemical precipitation.

The general east–west trend of the land mass to the north was paralleled by an elongated sedimentary prism in the form of a southwards-sloping carbonate ramp (Figure 8). The sediments deposited on the ramp are mainly of shallow-water facies; epeirogenic movement, either uplift to the north or depression to the south, and eustatic effects played an important part in determining sediment type and distribution.

Over most of the area, deposition was on broad flats submerged to only a few metres; thus minor changes in sea level or in the rate of elevation or of depression caused large effects. Sometimes, large areas of the carbonate ramp were exposed and eroded, or inundated by clastic material derived from the St George's Land–Brabant massif.

Interpretation of palaeobathymetric facies has to be undertaken with caution since similar rock types can be deposited in more than one environment, and different rock types can be deposited at the same water depth. Furthermore, material accumulated in one environment can be swept into another to its final depositional resting place. Nevertheless, after making allowances for these complicating factors, some generalisations can usefully be made. The shallowest facies are considered to be the calcite- and dolomite-mudstones ('chinastones' of earlier accounts). Such rocks are thought to represent deposition under peritidal conditions. Where they include abundant stromatolitic algal mats and little other sign of life, they are interpreted as supratidal or intertidal in origin. Similar rocks with a very sparse shelly fauna, though locally with immense numbers of a single brachiopod species, have been interpreted as lagoonal deposits. Nowadays, this term is restricted to deposits laid down behind an offshore barrier, and it has been suggested (Wilson et al., 1988) that the oolite shoals that were forming contem-

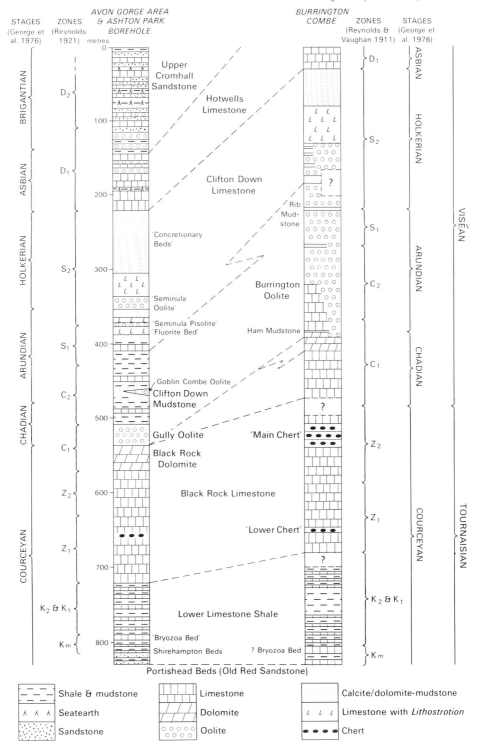

Figure 7 Comparative vertical sections of the Carboniferous Limestone in the Avon Gorge area, Bristol and at Burrington Combe, Mendip Hills to illustrate the relationship between the lithostratigraphical and chronostratigraphical classifications of the rocks.

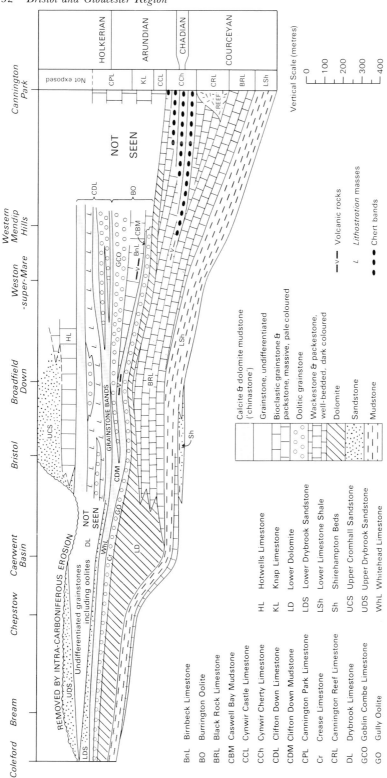

Figure 8 Generalised horizontal section to illustrate facies and thickness changes in the Carboniferous Limestone between the Forest of Dean and Cannington Park.

For location of section see Figure 6. The Dinantian stages in the right-hand column refer only to Cannington Park; for the relationship of the stages to the successions elsewhere refer to Figure 7 and 9.

BnL Birnbeck Limestone
BO   Burrington Oolite
BRL Black Rock Limestone
CBM Caswell Bay Mudstone
CCL Cynwir Castle Limestone
CCh Cynwir Cherty Limestone
CDL Clifton Down Limestone
CDM Clifton Down Mudstone
CPL Cannington Park Limestone
Cr   Crease Limestone
CRL Cannington Reef Limestone
DL   Drybrook Limestone
GCO Goblin Combe Limestone
GO   Gully Oolite

HL   Hotwells Limestone
KL   Knap Limestone
LD   Lower Dolomite
LDS Lower Drybrook Sandstone
LSh Lower Limestone Shale
Sh   Shirehampton Beds
UCS Upper Cromhall Sandstone
UDS Upper Drybrook Sandstone
WhL Whitehead Limestone

poraneously farther south provided such a barrier (e.g. the Burrington Oolite of the Mendip area).

The carbonate mudstones are extensively recrystallised and opinion remains divided as to how much of the lime mud represents finely comminuted organic mainly algal, debris and how much was chemically precipitated in the very shallow warm water. The dolomite in these and other rocks may be primary or secondary in origin, having been formed as a result of conditions of intense evaporation of either groundwater or very shallow sea water.

Another distinctive facies is oolite, which represents chemical precipitates form- ed in strongly agitated shallow water, probably no more than 5 m in depth. Such high energy conditions occur, for instance, in channels and on shoals, bars and underwater deltas, both within and outside the lagoonal areas. Cross-bedding is often conspicuous in these rocks, but the macrofauna is sparse due to the unstable substrate. Associated with the oolites, grainstones formed of micritised pellets and various intraclasts are widespread and probably represent the winnowed remnants of the muddier sediments. More open-sea or deeper-water conditions are re- presented by massive, pale grey, bioclastic limestones ranging from well-sorted grainstone, often finely cross-bedded, to less well-sorted packstone, often strongly bioturbated. The energy regimes range from high to medium. Crinoidal debris is ubiquitous, and fossil frequency and degree of sorting are inversely related. Col- onial and solitary corals, and thick-shelled brachiopods are characteristic. Deeper water conditions are thought to be represented by the well- bedded, dark grey to almost black, bioclastic limestone predominating in the Black Rock Limestone. This is a poorly sorted wackestone with an appreciable carbonate mud matrix, and is usually highly fossiliferous. The limestone may be bituminous and include chert in layers and nodules.

The deepest water carbonate facies, and hence the farthest downslope from the shoreline, is represented by the Waulsortian 'reefs' or bioherms (steep-sided lime- mud mounds) (Figure 8). Lees and Miller (1985) have suggested that the pioneer faunal community responsible for the establishment of this facies lived under aphotic conditions below wave base, at a depth of over 200 m. The pioneer com- munity was replaced by successive faunal and floral communities as carbonate was generated and accumulated, reflecting the growth of the bioherm above the sur- rounding sea floor. A belt of Waulsortian facies may contain many bioherms at different stages of growth. Although these bioherms have often been termed reefs, they lack any obvious skeletal framework and hence are not true reefs. The term Waulsortian, adopted from Belgium where this facies is well exposed, is used, therefore, to distinguish this unique Dinantian facies from other types of bioherm. Although this facies has only recently been described in the region from the Knap Farm Borehole at Cannington Park (see below), it provides valuable additional evidence for the presence, during an appreciable part of Tournaisian times, of a belt of Waulsortian reefs stretching in an approximate east–west direction be- tween Dinant in Belgium to the east and southern Ireland to the west.

The basin margin is marked west of Taunton by a belt of limestone turbidites and shales, the former presumably derived from the shelf carbonates. If this evidence to the west of the region is combined with seismic evidence (Figure 16) a short distance to the east of it, the basin margin can be predicted to trend at depth across the region more or less at the latitude occupied by Yeovil. Beyond this, to the south, within the sediment-starved basinal areas, the early Carboniferous is represented by an attenuated sequence of dark coloured, non-calcareous mudstones with chert and a mainly planktonic fauna including conodonts, gonia-tites and radiolaria.

STRATIGRAPHY

In the following account the Dinantian rocks are described in ascending strati-graphical order for the whole region, except for those at Cannington Park, which are treated separately.

Lower Limestone Shale

The Lower Limestone Shale is predominantly shaly, but limestone beds are typic-ally present in varying proportions. The formation can be divided into two une-qual parts, the lower part being characterised by coarse bioclastic and oolitic limestones, though mudstones and sandstones of Old Red Sandstone facies occur toward the base. Typically, the lowest part has a restricted marine- or brackish-water fauna with *Lingula*, small calcareous brachiopods, gastropods, bivalves in-cluding *Modiolus*, ostracods, calcareous algae, bryozoans, conodonts, serpulids and fish. The bioclastic limestones are commonly made up of worn crinoid columnals, but bryozoa debris may locally be abundant. Both these and the oolitic rocks com-monly show ripple-marking, current-bedding and various scour features. The limestones may be locally reddened by impregnation with hematite, either throughout the whole rock or just in the skeletal debris.

The upper part of the Lower Limestone Shale consists of greenish grey shale with interbedded dark grey to almost black, crinoidal limestone, which reflect a more open-sea environment. The limestone is often very fossiliferous, with a rich shelly fauna similar to that of the overlying Black Rock Limestone, though it is poor in corals. Passage into the Black Rock Limestone occurs with the upwards in-crease in the proportion of limestone.

In the Bristol area the lowest part of the formation is locally designated the Shirehampton Beds, with the 'Bryozoa Bed' at the top. The latter is the best known and most persistent of the reddened coarse crinoidal limestones and is overlain by the 'Palate Bed', a thin bone-bed that marks a nonsequence and the change to a dark shale sequence above. In the Mendips the Lower Limestone Shale is comparatively more shaly in its lower part, and the vertical transition from Old Red Sandstone is rapid. Nevertheless, at Burrington Combe a discontinuous reddened crinoidal limestone, some distance above the base, with a fish-bearing shale containing phosphatic nodules immediately above, invites comparison with the Bristol sequence.

The Lower Limestone Shale ranges in thickness from 150 m in the western Mendips to around 40 m on the west side of the Forest of Dean, with an inter-mediate figure (100 to 110 m) at Bristol and Portishead. This northwards attenua-tion is irregular and may have been affected by local epeirogenic movements (p.66).

The main exposures are quarry sections in the lower limestone facies. In the Mendips the railway cutting at Maesbury provides a more or less continuous sec-tion through the upper two-thirds of the succession, including the contact with the Black Rock Limestone.

Black Rock Limestone

This is the thickest and most consistent formation within the Carboniferous Limestone of the district. Most of the sequence comprises dark grey to almost black, well-bedded, poorly sorted crinoidal packstones and wackestones. Fossils are abundant, except where they have been obliterated by secondary dolomitisation, and have enabled ready subdivision into zonal assemblages based mainly on the conodonts, corals and foraminifera. In the Mendips the Black Rock Limestone thickens from west to east from about 250 to 370 m. When traced northwards, the thickness decreases markedly and the faunal evidence demonstrates that this is partly due to the removal or non-deposition of the lower part of the Chadian sequence (Figure 7). At Bristol the formation is about 150 m thick, including about 30 m of dolomite at the top, decreasing to between 105 m and 120 m in the north of the Bristol Coalfield and to as little as 70 m in the northern part of the Forest of Dean. There is a concomitant northwards increase in the dolomitisation of the upper part of the succession. In parts of the northern rim of the coalfield and everywhere west of the River Severn, the whole formation has been converted to dolomite. West of the Severn the dolomite facies is known as the Lower Dolomite and east of the Severn, the Black Rock Dolomite.

The Black Rock Limestone is most complete in the Mendips and is best examined in Burrington Combe, where detailed collecting has led to the establishment of faunal ranges throughout its thickness (Figure 7). Here, on the evidence of the foraminifera and conodonts, the uppermost one-third of the formation is of Chadian age. The Courceyan strata below are marked by a strong development of chert in bands and nodules, the so-called 'Main Chert'. In the upper part of the Black Rock Limestone, the higher beds tend to be paler in colour and better sorted than those below, and are packstones rather than wackestones. This is most marked in the eastern Mendips where the formation is thickest and includes strata younger than elsewhere. The best documented section here is east of Leigh-on-Mendip (Butler, 1973), where the Main Chert is 89 m thick. It is overlain by 44 m of dark wackestone facies rocks, above which there are 92 m of grey packstone-type crinoidal limestones with Arundian fossils in the upper part.

The early Courceyan part of the Black Rock Limestone is characterised by scattered zaphrentoid corals such as *Zaphrentites delanouei* and *Sychnoelasma* ['*Zaphrentis*'] *clevedonensis*, but is dominated by brachiopods including *Leptaena analoga, Rugosochonetes vaughani, Schellwienella* spp., *Schizophoria* spp. and *Unispirifer tornacensis*, which continue up to the late Courceyan. The late Courceyan part is marked by the incoming of a rich coral fauna including *Caninia cornucopiae, Caninophyllum patulum, Cyathaxonia cornu, Cyathoclisia tabernaculum* and *Synchnoelasma konincki*. The Chadian part is distinguished by the incoming of the large caninoid coral *Siphonophyllia cylindrica*. The brachiopods differ from those below, and chonetids and pustulose productoids become increasingly important.

The Black Rock Limestone, including the dolomitic facies, is widely quarried, especially west of the River Severn and in the more northerly area, east of the Severn.

CLIFTON DOWN GROUP (excluding the arenaceous facies)

Between the relatively homogeneous Black Rock Limestone and the Hotwells Group, the Clifton Down Group comprises two to six formations of varied facies, apparently deposited under peritidal or lagoonal conditions, in shallow or ex-tremely shallow water.

The local successions are shown in Figures 7 and 8, and their relationships to the older and newer Dinantian classifications are shown in Figure 7. They may be grouped into three rather indistinctly defined facies units that form wide belts trending in a general east – west direction and grading into each other in a north – south direction. These belts are named for purposes of description, from north to south, as the Forest of Dean, the Bristol and the Mendip regions respectively. The Forest of Dean region and the northern part of the Bristol Coalfield are charac-terised by the strong development of arenaceous facies, the Bristol region by peri-tidal carbonate mud sequences and the Mendip region by an appreciable overall thickening of the sequences, which are dominated by oolitic and bioclastic grain-stones and which formed as carbonate barrier shoals.

Gully Oolite

The Gully Oolite, known as the Crease Limestone west of the River Severn, is the lowest, most distinctive and widespread formation of the group. It is 20 to 30 m thick in the north-west and 35 to 40 m on Broadfield Down and in the Weston area in the south. Typically, it comprises pale grey, medium- to fine-grained, cross-bedded oolite, which is white-weathering and exceedingly massive, with strong vertical joints. Macrofossils are rare; Vaughan's name of 'Caninia Oolite' refers to its zonal position in his scheme. The lowest part commonly consists of pale grey, well-sorted, crinoidal limestones, known in the Bristol area as the 'Sub-Oolite Bed', which may contain an abundant brachiopod fauna of mixed Tournaisian and Viséan aspect, including chonetoids and large orthotetoids. The 'Sub-Oolite Bed' ranges from less than one metre to 7 m thick, but exceptionally, in the Tytherington area north of Bristol, it attains 21 m; the overlying oolite is up to 43 m thick. The diagnostic Chadian brachiopod *Levitusia humerosa,* common in northern England, has been found only near the base of the Gully Oolite, at Mid-dle Hope, north of Weston-super-Mare, but conodont faunas recovered from the Gully Oolite are consistent with a Chadian age.

The dolomitisation that affects the underlying Black Rock Limestone commonly extends upwards into the Gully Oolite without affecting the massive nature of the beds. At the southern limit of its outcrop, on Brean Down, it includes thick lenses of coarse crinoidal limestone with sparse coral faunas, but elsewhere in the Men-dips it cannot be separately distinguised from the overlying massive oolitic and bioclastic limestones that comprise the Vallis Limestone and the Burrington Oolite.

Clifton Down Mudstone

The Clifton Down Mudstone, or the Whitehead Limestone as it is known in the Forest of Dean, consists of thinly bedded, pale grey calcite and dolomite mudstones, commonly called chinastones, with thin, interbedded, dark grey and brown shales. The calcite and dolomite mudstones weather whitish grey or buff. They are poorly fossiliferous except for some foraminifera, stromatolitic algae and occasional serpulid worms. The formation is not present in the Mendips, apart from Brean Down, where it comprises 1.5 m of dolomitic mudstone and shale. From Bristol northwards it ranges in thickness from about 30 to 60 m.

The beds abruptly overlie the Gully Oolite and the boundary, which has been referred to as 'the mid-Avonian break', is one of the most distinctive in this and adjacent regions. In the Forest of Dean, where the junction is locally very irregular, differences in thickness in the Gully Oolite are attributed to erosion at this level. Farther south, for example in the Weston area, there is evidence of karstification and micritisation of the underlying strata, indicating subaerial exposure of the Gully Oolite before deposition of the Clifton Down Mudstone. Current opinion, however, favours the view that no great interval of time is represented by the erosional break, and the implied changes of sea level need not have been very great (p.29).

In the Bristol – Wick area intercalations of massive oolitic and crinoidal limestone, of which the Goblin Combe Oolite is the most important, are present in the higher parts of the formation. On Flat Holm, thick bioclastic limestones with some half dozen intercalations of the Clifton Down Mudstone facies have been given the name Flat Holm Limestone Member. There, the thickness of the beds containing the mudstone facies is about 35m, including a basal calcite mudstone – mudstone sequence with stromatolites, which has been separately named the Caswell Bay Mudstone after its type locality in Gower (George et al., 1976). Farther south, in the Weston area, the sequence above the Caswell Bay Mudstone comprises grey crinoidal limestones, to which the name Birnbeck Limestone has been applied. It is usually distinguished by a lower very massive, finely cross-bedded and sometimes oolitic facies, and an upper well-bedded, bioturbated facies. The lower facies is relatively poor in macrofossils apart from *Palaeosmilia murchisoni*, but the upper facies is fossiliferous and, in particular, is distinguished by the large keeled chonetoid *Delepinea carinata*, which is diagnostic of the Arundian Stage.

The Goblin Combe Oolite thickens southwards from the Bristol and Wick area at the expense of the mudstone facies. At its type locality at Broadfield Down, the oolite is 38m thick, but its thickest development is on the coast at Weston-super-Mare where it attains 70 m. It is distinguished from the Gully Oolite by its darker colour and coarser texture, and by the abundance of crinoidal debris. Scattered fossils include *Palaeosmilia murchisoni* and bellerophontid gastropods, which are very abundant in the Failand – Clevedon area. In the Weston – Bleadon area, the basal part of the Goblin Combe Oolite is marked by a single bed of pale grey, well-sorted, crinoidal limestone, 18 m thick on Steep Holm, and strongly contrasting with the well-bedded Birnbeck Limestone below.

Burrington Oolite

In the Mendips, apart from the small area between Bleadon and the coast, the Gully Oolite, Birnbeck Limestone and Goblin Combe Oolite of the Bristol – Broadfield Down – Weston area come together to form the Burrington Oolite. The individual components of the formation remain generally recognisable, but cannot readily be mapped out in the absence of any distinctive interbedded marker units such as the mudstone facies in the Clifton Down Mudstone. As the formation is traced southwards and eastwards towards the eastern Mendips, the non-oolitic facies become increasingly dominant and the term Vallis Limestone is used to describe the largely bioclastic sequence.

Clifton Down Limestone

Over the greater part of the district the uppermost formation of the Clifton Down Group is the Clifton Down Limestone, known as the Drybrook Limestone west of the Severn. Its type section is in the Avon Gorge (Figure 7), which illustrates both the general sequence and the variable nature of the rocks. The lowest part consists of well-bedded calcite and dolomite mudstones, with common stromatolites similar to those in the underlying Clifton Down Mudstone which, like them, indicate deposition in a very shallow-water, peritidal or lagoonal environment. Then follow rather poorly sorted, bioclastic limestones characterised by the cerioid species of the coral *Lithostrotion*. The first appearance of these characterises the base of the Holkerian Stage. This facies represents more open-sea conditions, though the recurrence of peritidal conditions from time to time is shown by the presence of the algal 'Seminula Pisolite' and beds of calcite mudstone with *Composita*. The Seminula Oolite occurs in the upper part of the sequence and may represent a channel or bar deposit. Finally, peritidal conditions returned with the upper thick calcite mudstone–algal sequence known as the 'Concretionary Beds'. A similar sequence occurs in the Wick Inlier east of Bristol (Murray and Wright, 1971).

This general succession can be traced southwards from Bristol where oolites become increasingly dominant in the lower part; silicified *Lithostrotion* masses and chert nodules and layers become prominent in the middle part, and algal growths become more conspicuous in the upper part. Thick lenticular masses of cross-bedded oolite, similar to the Seminula Oolite of the Avon Gorge, occur locally, but are not confined to any one horizon. Thus, the Cheddar Oolite occurs near the base of the sequence in the area after which it is named and the Brockley Oolite on Broadfield Down directly underlies the uppermost group of calcite mudstone beds.

Over most of the Mendips, calcite and dolomite mudstones form no more than 20 to 30 per cent of the succession below the *Lithostrotion*-bearing beds. The base of the Clifton Down Limestone is somewhat arbitrarily taken at the base of the lowest mappable calcite mudstone occurrence. In the Wells area, the lower calcite mudstones are apparently not present and the associated, mainly oolitic rocks are indistinguishable from and grouped with the underlying Burrington Oolite. At Burrington Combe, isolated calcite–mudstone beds, such as the Rib Mudstone (George et al., 1976), occur in the Burrington Oolite, but are too thin to map; here, the base of the Clifton Down Limestone is drawn at the base of the lowest thick unit of calcite mudstones above. At Holwell, in the eastern Mendips, the predominantly oolitic rocks in the lowest part of the Clifton Down Limestone are replaced by crinoidal bioclastic limestones, a change which indicates more open-sea conditions in that direction.

Northwards from Bristol the local facies variations in the Clifton Down Limestone are less well known. At Chipping Sodbury, about 11 m of stromatolite-bearing rocks are seen at the bottom of the sequence; they are overlain by c.103 m of thick *Lithostrotion*-bearing beds (Murray and Wright, 1971) and then by c.16 m of calcite mudstone with stromatolites. The most striking difference in the succession here is the presence of wedges of arenaceous facies that thicken northwards at the expense of the intervening limestone (see below). West of the River Severn, a wholly arenaceous sequence north of Cinderford in the Forest of Dean is represented by the Drybrook Sandstone; farther south, it includes a high proportion of oolites.

The Clifton Down Limestone approximates to the Holkerian Stage, and is characterised by an assemblage which includes the brachiopods *Davidsonina carbonaria*, *Composita ficoidea*, *Linoprotonia corrugatohemispherica* and the corals *Axophyllum vaughani* and *Lithostrotion aranea*. Holkerian foraminifera, including *Holkeria avonensis* and *Pujarkovella nibelis*, have been recorded from the Clifton Down Limestone in the Bristol area.

HOTWELLS GROUP (excluding the arenaceous facies)

Hotwells Limestone

The calcareous facies of this group is known as the Hotwells Limestone and, although the type locality is in the Avon Gorge, it reaches its maximum development on Broadfield Down and the Mendips. Typically, the formation comprises massive, grey, oolitic and crinoidal limestones with an abundant fauna of corals and thick-shelled brachiopods. It marks an abrupt change in depositional environment from the predominantly low energy peritidal conditions of the upper part of the Clifton Down Limestone to an open-shelf marine environment characterised by turbulent, high-energy conditions. On faunal grounds, it appears that in contrast to northern England there is a nonsequence at the base of the Hotwells Limestone in this region. The thickest dominantly calcareous sequences are present at Winford (225 m) in the south-eastern part of Broadfield Down and along the northern margin of the Mendips between Ubley and Mells (180 to 225 m). In these areas, above a relatively uniform sequence in the lower third of the formation, the beds display evidence of the cyclical sedimentation that is so marked in the areas farther north (see below). The cycles are characterised by thick massive limestones alternating with thin black, commonly carbonaceous shales. Beneath the shales there may be calcareous rubbly beds of pedogenic aspect. On the south side of the Mendips, the full thickness of the Hotwells Limestone, reduced to 125 m, is seen only at Ebbor, north-west of Wells. Details of the succession are unknown and it is possible there is attenuation due to pre-Namurian erosion (p.66). Northwards from the Mendips, the limestone facies increasingly pass into rhythmic, predominantly arenaceous sequences of sandy limestone, sandstone and shale, which include seatearth and very thin coals. At Bristol, the top half of the Hotwells Limestone has been similarly replaced (see below), and along the northern rim of the coalfield the Hotwells Limestone is largely unrecognisable. The formation is not known anywhere west of the River Severn.

Chronostratigraphically, the Hotwells Limestone spans the Asbian and the Brigantian stages. The Asbian is characterised by a rich fauna which includes the corals *Lithostrotion pauciradiale, L.junceum* and *Palaeosmilia murchisoni*, and the productoids *Gigantoproductus* spp. and *Linoprotonia hemispherica*. The contact with the Brigantian is poorly defined, partly because of the presence of unfossiliferous arenaceous intercalations at about this level and partly because the characteristic Brigantian corals appear to be confined to discrete horizons (cf. Rownham Hill Coral Bed; George et al., 1976, p.17). The corals include *Lonsdaleia floriformis, Nemistium edmondsi, Orionastraea* spp. and *Palaeosmilia regia*. These occurrences have been recognised over a wide area extending from north of the Bristol Coalfield, eastwards to Wick and southwards to Wrington and Compton Martin.

ARENACEOUS FACIES IN THE CLIFTON DOWN AND HOTWELLS GROUPS

The collective name given to the arenaceous facies within the Clifton Down and Hotwells groups is the Cromhall Sandstone east of the River Severn and the Drybrook Sandstone west of the River Severn. These sandstone successions are thickest in the north; southwards they split into several units separated by calcareous sequences.

The Lower Cromhall Sandstone, which separates the Clifton Down Mudstone from the Clifton Down Limestone, extends as far south as Chipping Sodbury and is equivalent to the Lower Drybrook Sandstone west of the River Severn. The Middle Cromhall Sandstone separates the Clifton Down and Hotwells limestones and extends south to beyond Wick. The Upper Cromhall Sandstone replaces the Hotwells Limestone from the top downwards in a northerly direction. This is the thickest and most extensive of the arenaceous developments in the Dinantian sequence and at its southern limit reaches the Mendips at Compton Martin. West of the River Severn, the Upper Drybrook Sandstone rests on the Drybrook Limestone but, in the absence of any recognisable Hotwells Limestone above, it has not been possible to determine its exact equivalence to the sequence on the east side of the river. The Lower and Upper Drybrook sandstones come together at about the middle of the Forest of Dean due to the northwards wedging out of the Drybrook Limestone.

The best documented sequence of the Upper Cromhall Sandstone is in the Ashton Park Borehole (Kellaway, 1967), a short distance south of the Avon Gorge. There, the Hotwells Group is 221 m thick and can be divided into fourteen sedimentary cycles. Those typical of the Hotwells Limestone comprise, in upwards succession, black mudstone – limestone – sandy seatearth, and those typical of the Cromhall Sandstone comprise limestone – banded sandy and silty beds with thin shales-seatearth. The latter cyclothems are similar to those of the Yoredale facies of northern England and the change in character of the sediments between the two types of cyclothem indicates a change from the predominantly carbonate platform conditions of the Hotwells Limestone to the deltaic conditions of the Upper Cromhall Sandstone. More restricted marine conditions are suggested by the replacement of crinoidal limestone by oolitic limestone in the later cyclothems. The Mollusca Band at Wick, and the Tanhouse Limestone in the country to the north, are thin limestones in the uppermost part of the Upper Cromhall Sandstone and are useful markers for separating this formation from the overlying Quartzitic Sandstone Formation.

The Hotwells Group is thickest at the northern end of the Bristol Coalfield, at Yate, where the Upper Cromhall Sandstone is 250 m thick and overlies some 70 m of Hotwells Limestone, which includes beds of Middle Cromhall Sandstone in the middle and at the base. The succession at Wick is intermediate between this and Bristol (Figure 9). The thickest Drybrook Sandstone in the Forest of Dean is only 105 m, apparently due to intra-Carboniferous erosion (p.66). The Hotwells Group may also be represented amongst relict brecciated beds along the line of the Lower Severn Axis (pp. 47,66).

CANNINGTON PARK

The Carboniferous Limestone inlier of Cannington Park, which lies 5 km northwest of Bridgwater, is the carbonate platform locality nearest to the Lower Carboniferous basinal sequences of west Somerset and Devon. Structurally, it forms part of the northern limb of the main Quantocks Anticline. Knowledge of the succession has been greatly increased by the BGS Knap Farm Borehole (Whittaker and Scrivener, 1982). The borehole succession is as follows:

	m Approximate thickness (corrected for dip)	m Depth
CARBONIFEROUS LIMESTONE		
Cannington Park Limestone	73	93.2
Knap limestone	195	249.55
Cynwir Castle Limestone	278	351.21
Cynwir Cherty Limestone	446	557.68
Cannington Reef Limestone	626	778.41
Black Rock Limestone	776	966.35
Lower Limestone Shale	890	1105.91
DEVONIAN SANDSTONE		seen to 1153.00

The borehole is situated on the southern edge of the inlier; thus at least 150 m must be added for the thickness of the Cannington Park Limestone at outcrop, to give a total thickness of the Carboniferous Limestone in excess of 1000 m. This is considerably greater than that of the corresponding strata in the Mendips, but is comparable with that known in south Pembrokeshire. The dips in the core range from 30° to 40°. The limits of the stages are given in Figure 8; the youngest out-cropping strata belong to the Holkerian Stage.

In terms of the Mendip sequences, the top three formations at Cannington Park are equivalent to the Burrington Oolite in the western and central Mendips, and the Vallis Limestone in the eastern Mendips (Figure 8). At Cannington Park, these formations comprise pale grey, bioclastic and oolitic, coral-bearing rocks, which extend up into the Holkerian; they correspond to the generally finer-grained, more variable Clifton Down Limestone and Burrington Oolite in the areas to the north. Grains described as ooliths during field-logging of the Knap Farm borehole cores were subsequently shown by detailed petrographic work to be mostly well-rounded, micritised, crinoidal and other debris. It is probable, like-wise, that the oolitic content of the Burrington Oolite of the Mendips has been over-estimated.

The Cynwir Cherty Limestone is similar to the upper part of the Black Rock Limestone of the Mendips, except that the chert development in the Mendips is thinner. The extension of the Black Rock Limestone facies up into post-Chadian strata is paralleled by a similar development in the eastern Mendips (p.35). At Cannington Park, it is noteworthy that the evidence for appreciable shallowing of the sea in post-Courceyan times is first seen in the early Arundian, and that there is no obvious sedimentary evidence for the intra-Chadian sea level lowering that so strongly affected sedimentation to the north. Thereafter, it appears that the rate of sedimentation was greater than the rate of subsidence and, in consequence, a shallowing of the sea took place.

IGNEOUS ROCKS

Minor submarine volcanic activity is recorded at two stratigraphical levels in the Carboniferous Limestone of the district. The earlier episode occurred in the later Courceyan rocks of the Middle Hope area north of Weston-super-Mare, where up to 32 m of tuffs with some interbedded tuffaceous limestones are exposed on the coast (Plate 6A). The tuffs show slight signs of reworking and may be adjacent to

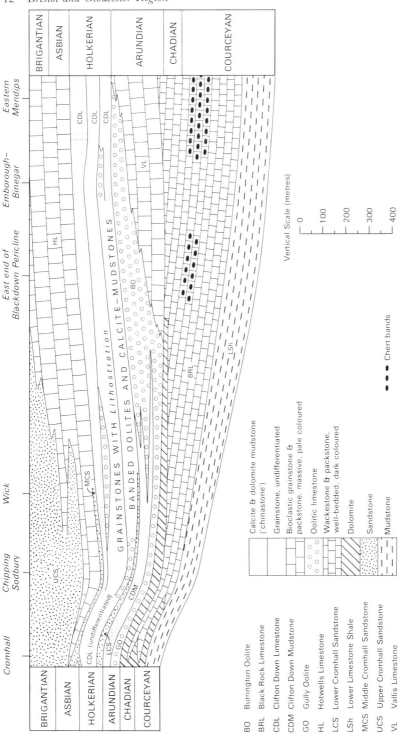

Figure 9   Generalised horizontal section to illustrate facies and thickness changes in the Carboniferous Limestone between Cromhall and the eastern Mendips. For location of section see Figure 6.

Plate 6 Lower Carboniferous (Dinantian) volcanic rocks at Weston-super-Mare.
A. Volcanic tuff at Middle Hope (A11767).
B. Pillow lava at Spring Cove (A11788).

the volcanic source. A single flow of basaltic pillow lava about 4 m thick occurs near the top of the sequence. The tuffs rest on the 'Main Chert' of the Black Rock Limestone at about the (local) middle of the *Caninophyllum patulum* Biozone of late Courceyan age (Figure 8). At the same stratigraphical level as the Middle Hope volcanics, a thin basalt 'lava', formerly exposed in a quarry adjacent to the railway at Uphill, is now considered on mineralogical evidence to be a sill.

Later activity, in the early part of the Arundian, is recorded at four localities south-west of Bristol, stratigraphically about 15 m above the top of the Gully Oolite (Figure 8). The best exposures are on the coast at Spring Cove, Weston-super-Mare (Plate 6B), within the Birnbeck Limestone. Here, about 12 m of reddened tuffaceous limestone are overlain by a 15 m-thick flow of basaltic pillow lava, with a further 8 to 9 m of reddened and tuffaceous limestones lying above this. The tuffaceous material has been reworked and the flow contains much partly assimilated and baked limestone debris. Some 3 km inland, the flow locally reaches a maximum thickness of 35 m. About 13 km further to the east, basaltic lava and tuff crop out in Goblin Combe, on the south side of Broadfield Down, and extend over 3 km eastwards along the strike. The volcanic sequence is about 12 m thick and lies entirely within the Clifton Down Mudstone, which is the lateral equivalent of the Birnbeck Limestone (p.41). The remaining two localities, at Tickenham and Cadbury Camp, east of Clevedon, comprise small isolated outcrops of basalt within the Clifton Down Mudstone. The relatively small thickness of the lava flows, combined with their relatively wide area of occurrence (some 250 km²), suggest that there were several small volcanic vents. Re-examination of the petrology of the lavas (Whittaker and Green, 1983) shows that they are all alkaline olivine basalts which have undergone various degrees of autometasomatism by alkali-rich residual fluids.

West of the River Severn, two small intrusions of monchiquite have been dated as post-early Dinantian. One, at Guest House, 6.5 km south-east of Usk, appears to be a small volcanic neck at least 46 m across and mainly filled with agglomerate. This consists of blocks of decomposed monchiquite, along with rocks that can be matched with limestones in the Lower Limestone Shale and dolomites in the Lower Dolomite. The other occurrence is of a small monchiquite dyke intrusive into Lower Devonian nearer Usk and just beyond the limits of the region. It is not known whether they are contemporaneous with the Bristol–Weston occurrences.

6   Upper Carboniferous (Namurian)

In the last century, the lithostratigraphical subdivision of the Upper Carboniferous sequence into Millstone Grit and Coal Measures was found to be applicable to virtually the whole of the British Isles. Since the 1920s, however, biostratigraphical research on the Millstone Grit and productive Coal Measures has shown that the lithostratigraphical boundary between them is of different ages in different areas. The term Millstone Grit is now used informally and the lower division of the Upper Carboniferous, defined largely in terms of subzones, zones and stages by means of the rapidly evolving goniatite faunas, has been redefined chronostratigraphically as the Namurian Series (Ramsbottom et al., 1978).

The St George's Land – Brabant landmass persisted from Dinantian into Namurian times. The main area of marine sedimentation was south of the Bristol Channel. Farther north, in South Wales and the Bristol area, there appear to have been separate basins in Namurian times with only intermittent marine connections via the deeper sea area to the south. Marine influences were minimal in the Bristol basin, where sedimentation was predominantly in a deltaic and swamp environment, and the detailed zonal subdivision based on goniatites cannot be applied here.

QUARTZITIC SANDSTONE FORMATION

In the Bristol area, Namurian rocks constitute the Quartzitic Sandstone Formation (Kellaway and Welch, 1955). The main outcrop occurs on the northern and north-eastern rim of the Bristol Coalfield between Chipping Sodbury and Cromhall (Figure 6), where the succession attains its maximum known thickness of about 300 m. Elsewhere, the outcrops are mainly small inliers, partly or wholly surrounded by Triassic rocks. At Bristol, the Quartzitic Sandstone gives rise to high ground in the Clifton – Tyndall's Park area; hence the local name Brandon Hill Grit, which was formerly used for these rocks. In the Wick inlier, farther south, the succession is about 180 m thick, and the Ashton Park Borehole near the southern end of the Avon Gorge proved a true thickness of 155 m. To the west of Bristol and across the River Severn, the formation is absent or strongly attenuated, due to intra-Carboniferous earth movements and erosion (p.66). In the Mendips, the thickness varies between 45 m and 65 m. The most southerly known occurrences of Namurian strata in the district are small inliers north-west of Bridgwater (see below).

BRISTOL AREA

The formation is very poorly exposed and detailed knowledge of the succession mainly derives from the Ashton Park Borehole and the Limekilns Lane Borehole at Yate; however, only in the former is the complete sequence seen. At Bristol and Winford, the base of the formation is marked by a distinctive development of chert and cherty mudstone, up to 15 m thick, with a marine fauna mainly of brachiopods and bivalves, but also including fish fragments and conodonts. At Winford, these beds yielded the only goniatite so far found in the formation, a *Eumorphoceras* of a type found in the earliest Namurian stage, the Pendleian (E1 zonal index). A much attenuated representative of the chert is present in part of the coalfield area north of Bristol, but it appears to be absent elsewhere. The top of the formation is taken at the base of the Subcrenatum Marine Band (Westphalian). Much of the succession, particularly in the upper part, is lithologically indistinguishable from the Coal Measures and was laid down under similar conditions.

Above the basal cherts, the formation comprises varying proportions of mudstone, seatearth and sandstone, with occasional marine mudstone layers and thin carbonaceous and coaly deposits. The rocks show much lateral variation. The sandstone beds commonly have erosive bases, and some may be pebbly or conglomeratic; a high proportion are very hard quartzite. Most of the pebbles are of white quartz, but also include chert, quartzite, siderite ironstone and mudstone. The sandstone beds are best developed at Bristol and north-eastwards to the western side of the coalfield around Tytherington and Cromhall; elsewhere, north-east and east of Bristol, the succession is more argillaceous. Thus, in the Ashton Park Borehole, arenaceous rocks account for three-fifths of the succession, of which two-thirds are quartzitic types; whereas, in the Limekilns Lane Borehole, in a similar thickness of strata, the corresponding proportions are only one-fifth and one half respectively. Some inferior quality coal seams have been worked locally near Barrow Gurney on the northern side of Broadfield Down, at Wick, and in the Tytherington–Cromhall area, where the crop of a coal, known as the Tapwell Bridge Seam, is shown on the Malmesbury (251) sheet.

The only biostratigraphical subdivision of the main mass of the formation has been by means of fossil plants. In the Ashton Park Borehole, the lower half of the succession contains plants having affinities with late Viséan to early Namurian (Pendleian–Arnsbergian) floras; in the upper half they have affinities with floras of the two highest Namurian stages (Marsdenian and Yeadonian) and the early Westphalian (see also Ramsbottom et al., 1978, p.16). Deposits of the Chokierian, Alportian and Kinderscoutian stages may not have been represented in this area.

SOUTHERN AREAS

On the north side of the Mendips, between Ashwick and Mells, the Quartzitic Sandstone forms a narrow outcrop on the northern limb of the Beacon Hill Pericline, where it is estimated to be about 45 m thick. The only exposures are in red-stained quartzitic sandstones. Apart from other small outcrops in the Mendips, the full thickness is only again seen at Ebbor Rocks, some 4 km north-west of Wells. Here, the Subcrenatum Marine Band overlies about 65 m of strata above the Hotwells Limestone. This thickness includes an unexposed gap of about 12 m at the base, apparently mainly in mudstone, possibly comparable to the lowest marine shales seen farther north. Although incompletely exposed, the bulk of the overlying strata appear to be quartzitic sandstone.

An interesting discovery of Namurian strata was made in the Cannington Park area, north-west of Bridgwater (Edmonds and Williams, 1985). Here, an inlier of Rodway Beds (now named Rodway Siltstones), hitherto considered to be Devonian in age, was proved by drilling to contain the goniatite *Gastrioceras cancellatum* indicative of the *G. cancellatum* Marine Band at the base of the Yeadonian Stage. A second inlier of Rodway Siltstones, nearby, is presumed also to be of Namurian age. The rocks are siltstone, sandstone and shale, too disturbed for any reliable estimates of thickness to be made.

WESTERN AREAS

Vestigial occurrences of possible Namurian strata in the coastal strip on the west side of the River Severn between Portskewett and Ifton have been described by various authors (Welch and Trotter, 1961). These deposits comprise masses of hard, partly quartzitic sandstone and soft shale, which infill steep-sided channels in the eroded top surface of the Drybrook Limestone (Dinantian) and connect with pipes and cavities in the underlying limestone. In the same area, a shaft sunk near the western end of the Severn Tunnel during its construction proved brecciated Carboniferous Limestone overlain by thin 'Millstone Grit' which was, in turn, unconformably overlain by Upper Coal Measures. More recently, exploratory boreholes at Portskewett have yielded palynological evidence for a Namurian (Yeadonian) age for shales overlying limestone there.

On the opposite side of the River Severn, other occurrences of lenticular masses of quartzitic sandstone, unconformably overlying brecciated Carboniferous Limestone of possible late Viséan age and unconformably overlain by Upper Coal Measures, have been recorded during tunnelling operations at Kings Weston, north-west Bristol, and at Portishead. Lenticular and piped masses of quartzitic sandstone are present at the top of the Clifton Down Limestone at Tynesfield, 1 km west of Wraxall. The strong unconformity and erosion associated with these vestigial remnants of proved and possible late Viséan-Namurian strata provide evidence for activity along the line of the Lower Severn Axis (p.66) during Carboniferous times.

7 Upper Carboniferous (Westphalian and Stephanian)

The Coal Measures in the district span the Westphalian and lowermost Stephanian series. Apart from the western edge of the concealed Oxfordshire Coalfield, which is just within the eastern limits of the district, the Coal Measures occur in the exposed Forest of Dean Coalfield to the west of the River Severn, and in the Bristol and Somerset coalfields to the east of the Severn (Figure 10). The latter are largely (73 per cent) concealed by Mesozoic rocks.

Post-Westphalian earth movements have led to the separation of the Bristol and Somerset coalfields into a number of structurally distinct areas previously referred to as basins (Figure 12). The Radstock and Pensford synclines, to the south, are jointly referred to as the Somerset Coalfield, and the Kingswood Anticline and Coalpit Heath Syncline, to the north, as the Bristol Coalfield. The Coal Measures of the Nailsea Syncline, though forming a structurally distinct domain, are in geological continuity beneath the Mesozoic cover with the Somerset Coalfield to the south and south-east. The Severn Coalfield, and the Clapton-in-Gordano and tiny Barrow Gurney inliers, which lie to the west of Bristol, are structurally separated from the main coalfield. The Coal Measures known to occur on the south side of the Mendips and in the Westbury (Wiltshire) Borehole are also presumed to be structural outliers of the main coalfield. The maximum preserved thickness of Coal Measures occurs in the Somerset Coalfield where it is between 2500 m and 2600 m.

CLASSIFICATION

Buckland and Conybeare (1824) were the first to divide the Coal Measures of the district into three major lithological groupings, namely the 'lower' and 'upper coal shales', separated by a thick division of grey sandstone termed the 'Pennant Grit'. Subsequent classifications of the Coal Measures built on this scheme mainly by further subdividing the upper and lower divisions on the basis of their contained coals. Subdivision of the Coal Measures has also been accomplished by using zonal schemes based on plants, nonmarine bivalves and miospores, but the basic classification depends on the recognition of marine horizons within the otherwise nonmarine sequences. The development and systematic application of such fossil-based schemes of classification, for example to correlation, depend for their success on geological examination of the strata as exploration proceeds. This only became standard practice following nationalisation of the coal industry in 1945, by which time the heyday of coal mining was past and new exploration work considerably reduced. Nevertheless, by the time the last pit closed in 1971, the accumulated

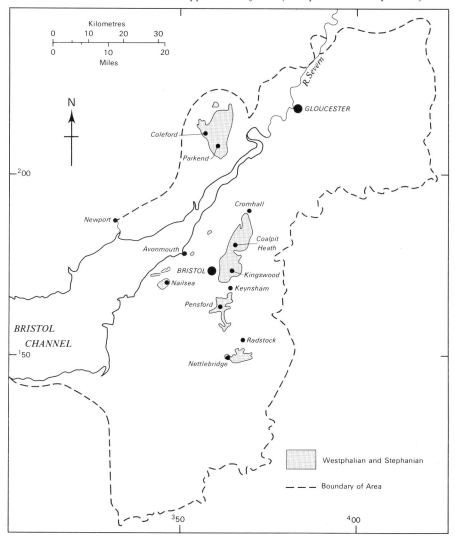

Figure 10 Outcrop of the Westphalian and Stephanian rocks in the region.

data enabled application of the standard British chronostratigraphical classification of the Coal Measures to the greater part of the district (Table 3).

This was systematised by the Upper Carboniferous Working Group of the Geological Society of London (Ramsbottom et al., 1978). Their recognition of the fourfold division of the Westphalian into stages labelled A,B,C and D has been formalised by Owens et al., (1985) who introduced the names Langsettian, Duckmantian and Bolsovian respectively for Westphalian A, B and C. A stratotype for the Westphalian D division has yet to be designated. The florally defined Westphalian E has now been abandoned and redefined as part of the Stephanian Series, and an intermediate stage, the Cantabrian, has been recognised between Westphalian D and the former Westphalian E. It is now thought

Table 3   Classification of the Coal Measures (after Ramsbottom et al., 1978). The spore zones are after Smith and Butterworth (1967).

Stages	Marine bands	Nonmarine bivalves		Spores	Divisions on Geological Survey maps
		Zones	Faunal belts		
Stephanian					
Cantabrian					
Westphalian D		*A. tenuis*		*Thymnospora obscura* (xi)	Upper Coal Measures
Westphalian C (Upper) [Bolsovian]	Cambriense*	*A. philipsii*		*Torrispora securis* (x)	
Westphalian C (Lower) [Bolsovian]	Aegiranum*	Upper *similis-pulchra*	*adams-hindi** *atra**	*Vestispora magna* (ix)	Middle Coal Measures
Westphalian B [Duckmantian]	Vanderbeckei*	Lower *similis-pulchra*	*caledonica** *phyrgiana* *ovum** *regularis*	*Dictyotriletes bireticulatus* viii)	
Westphalian A [Langsettian]		*A. modiolaris*	*cristagalli* *pseudorobusta**	*Schulzospora rara* (vii)	Lower Coal Measures
		C. communis	*bipennis* *torus* *proxima*	*Radiizonates aligerens* (vi)	
	Listeri*? Subcrenatum*	*C. lenisulcata*	*extenuata** *fallax-protea*	*Densosporites anulatus* (v)	

*Marine bands and nonmarine bivalve faunal belts recognised in the Bristol and Somerset coalfields

that the highest beds of the Forest of Dean Coalfield are probably Cantabrian in age and they are, therefore, the youngest Coal Measures strata in the district (Ramsbottom et al., 1978).

The major stage boundaries are, with the exception of that taken between Westphalian C (Bolsovian) and D, drawn at the bases of marine bands. Until the stratotype for the Westphalian D division has been designated, its base is taken at the incoming of the nonmarine bivalve *Anthraconauta tenuis* and miospores characteristic of the *Thymnospora obscura* (xi) miospore zone. Marine bands over the years have been given different names in different coalfields, but those included in Table 3, which are based on the principal diagnostic goniatites, are applicable throughout Britain (Ramsbottom et al., 1978, p.45). In the Bristol and Somerset coalfields the Subcrenatum, Vanderbeckei, Aegiranum and Cambriense marine bands have been known as the Ashton Vale, Harry Stoke, Crofts End and Winterbourne marine bands respectively.

The term 'Pennant', which has been used in many different ways in the past, was restricted by Ramsbottom et al. (1978) to the Forest of Dean Coalfield where, the 'Pennant' facies is now designated the Pennant Formation and the overlying beds the Supra-Pennant Formation. On the most recent Geological Survey maps of the Bristol–Somerset coalfields, the 'Pennant' facies is designated Pennant Measures and subdivided where possible into the Downend and Mangotsfield formations, whilst the overlying coal measure formations are grouped together as Supra-Pennant Measures.

COAL MINING

The working of coal in the Bristol, Somerset and Forest of Dean areas is of great antiquity and may date back to Roman times. Coal working in the Kingswood district of Bristol is mentioned in the Great Pipe Roll of 1223, but it was not until after Kingswood Chase had passed out of royal hands in 1594 that coal mining became a thriving industry. It flourished in the Bristol district until the early part of the present century, reaching its peak between 1870 and 1890. Coal mining in the royal Forest of Dean certainly goes back to before 1282, the working of the leases or 'gales' being carried on by the Free Miners.

In early times, coal was dug mainly for use by smiths and lime-burners, the public being prejudiced against its use as a domestic fuel. However, with the increasing shortage of wood in the towns in Tudor times, it became an essential fuel.

The early workings were sited in places where the coal cropped out and could be won by shallow excavations or by pits and levels. In the Forest of Dean the whole coalfield is exposed, so that workings developed in belts along the outcrops. In the Bristol and Somerset coalfields where only part of the field is exposed, the earliest workings comprised shallow pits whose development was limited by drainage and ventilation problems. In the Forest of Dean much coal was won by adit levels which provided a natural drainage.

The hand windlass or horse drum was the sole method of winding coal until about the middle of the eighteenth century when the steam engine came into use. With improved machinery, deeper mining became practicable and by about 1800 the average depth of the pits was around 150 m. The deepest shafts were at Braysdown (569 m) and Mendip (Strap) Pit, near Stratton on the Fosse (559 m).

Long-wall working, with a continuous or stepped face, became almost universal, an exception being the use of the pillar-and-stall method in the thick basal seam of the Coalpit Heath–Parkfield district. In the steep or vertical seams of the Nettlebridge and Vobster areas, methods resembling the stoping used in metal mines were adopted. The impending demise of the industry became clear in the late fifties due to the approaching exhaustion of readily winnable reserves, the impossibility of applying large-scale mechanisation methods to the thin coals and the difficult geological conditions generally prevailing in the area. With the exception of a few small drift mines worked by Free Miners in the Forest of Dean, the last mine was closed in 1973.

TYPES OF COAL

The coals of the Bristol–Somerset area and the Forest of Dean are essentially bituminous, the majority of them having strong coking properties. During recent times over 40 per cent of the output of the Bristol–Somerset area was used by the gas industry. In the Forest of Dean, the Pennant Formation provided gas-making and long-flame steam coals, whilst the Supra-Pennant Formation furnished house coals.

A factor that greatly contributed to the safety and ease of working in both coalfields was the almost complete absence of firedamp in the mines and, with a few exceptions, naked light working was possible.

CONDITIONS OF FORMATION OF THE COAL MEASURES

Sedimentation and uplift during the later part of Lower Carboniferous times finally converted most of the region into land. In a wide area east of the line of the Lower Severn Axis, the surface was marginally above the level of the sea, which lay far to the south. This area, into which the rivers from the higher ground discharged their sediments, was a vast swamp. Here flourished the great forests whose rotting debris accumulated as thick layers of peat, which ultimately formed the coal seams.

The climate of the period appears to have been warm, with high perennial rainfall, so that in the low-lying areas a wet substratum with a high water table was permanently maintained. This produced stagnant swamp conditions in which aerobic decay of the vegetable matter was arrested. In the more elevated regions bordering the swamp, where the water table lay below the ground surface, downward percolation of rain water supplied oxygen, which enabled complete bacterial decay of the vegetation and removal of the humic acids produced by fermentation. The resultant downward movement of the acid-charged waters hastened the decomposition of the underlying substrate and produced lateritic soils of a dominantly red colour, in contrast to the grey or black soils produced by the reduction of the ferric oxides in the anaerobic swamp substrates.

The higher ground along the Lower Severn Axis probably formed the western margin of the swamp: its eastern boundary may have been a ridge of pre-Carboniferous rocks coincident with the Malvern 'Axis', extending as far south as the Sodbury area (see Figure 11). Farther south, the older rocks along the 'Axis' plunged beneath the swamp which probably extended westward into Wiltshire.

The most striking feature about Coal Measure sedimentation is its cyclicity. The deposition of the organic debris that formed each coal seam was normally followed by flooding of the swamp, which on some occasions involved ingress of the sea. Mud and sand were then deposited until the swamp resumed its original level and, once again, vegetation grew. The soil, still preserving traces of rootlets, is represented by the underclay or seatearth that lies beneath a coal seam.

Swamp conditions persisted into mid-Westphalian C (late Bolsovian) times, with occasional temporary incursions of the sea caused by minor changes in sea level. Towards the end of this period, general uplift with folding and erosion of the surrounding areas was followed by a widespread change in sedimentation. Marine incursions ceased and in place of shales, fireclays and coal seams, a great thickness of deltaic sediment, comprising coarse-grained, grey, current-bedded, feldspathic, subgreywacke-type sandstone, known as the Pennant Formation (or Measures; see above) was laid down in a belt across the district and in South Wales.

In the Forest of Dean and north Bristol coalfields, beds of red measures, including conglomerates and seatearths without coals, immediately beneath the Pennant sandstones heralded the change in sedimentation. The onset of arenaceous sedi-

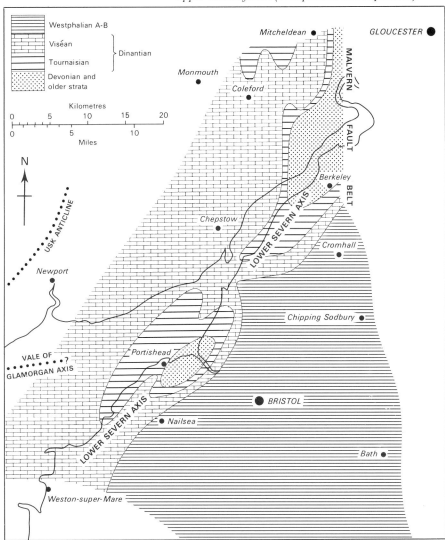

Figure 11 Sketch map showing the probable outcrops at the surface on which Pennant Measures (Westphalian C; Bolsovian) were deposited. Thin Namurian strata, themselves strongly unconformable on the underlying strata, have been ignored.

mentation was earliest in the south, both in the present district and in South Wales, and it is now generally considered that the sediment was derived from an uplifted landmass of Devonian rocks to the south, in the general latitude of the Bristol Channel. Sometimes there was local silting up of the basins, which allowed the establishment of swamp conditions, so that coal seams, usually of limited extent, were formed. In general, the cycles in the Pennant sandstones are thicker than in the measures above and below. The arenaceous sediments filled up all the low ground, creating deltas that spread over the surrounding higher ground, with resultant overstep onto older formations. Figure 11 shows diagrammatically the formations on which the arenaceous part of the Coal Measures (or more strictly beds of the *phillipsii* Biozone) is believed to rest. When the rivers had reached their base level, the supply of sand decreased, swampy conditions were re-established on the delta top and the Supra-Pennant Formation (or Measures) was laid down.

In contrast with most of the lower strata, the Supra-Pennant rocks include thick seatearths associated with red or barren measures containing poorly preserved plants and shells. The red measures may indicate arid or semiarid conditions, but the plant and shell fossils indicate that there was sufficient rainfall to support plants and provide freshwater conditions conducive to the existence of nonmarine bivalves. The absence of coal above the seatearth suggests a periodic lowering of the water table, permitting the rapid decay and erosion of the dead vegetation and oxidation of the substrate; but it cannot be ruled out that, in some cases, the red coloration of the measures may be due to a later, secondary cause.

PLANTS OF THE COAL MEASURES

The Upper Carboniferous was the age of the great arborescent lycopods and the pteridosperms whose descendants now occupy a humble place in today's floras, but which in Coal Measure times contributed to the equivalent of today's rain forests. The chief groups of plants were: Lycopodiales e.g. *Lepidodendron* and *Sigillaria*, which were tree-like 'clubmosses' up to 30 m in height; Equisetales e.g. *Annularia* and *Calamites*, which were tall 'horsetails' of similar height; Sphenophyllales, e.g. *Sphenophyllum,* which were slender and herbaceous, with leaves in whorls; Pteridosperms or 'seed ferns' of numerous and varied form which include most of the so-called Coal Measure 'ferns' such as *Neuropteris, Alethopteris, Linopteris, Mariopteris, Sphenopteris* and *Lonchopteris;* Filicales or 'tree ferns,' including pecopterids, e.g. *Asterotheca.*

The higher measures of Bristol, Somerset and the Forest of Dean contain some of the youngest fossiliferous Coal Measures of the British Isles. In these there is a great abundance of pecopterids, whilst *Calamites, Lepidodendron* and *Sigillaria,* so abundant in the lower measures, are greatly reduced in numbers.

FAUNA OF THE COAL MEASURES

The most interesting faunas occur in the marine bands which may show both lateral and vertical faunal variation (Calver, 1969) and which are attributed to open-marine conditions. Nearshore mudstones were dominated by the horny brachiopods *Lingula* and *Orbiculoidea.* They gave way farther offshore to a productoid facies dominated by a benthonic assemblage of calcareous and horny brachiopods, bivalves and gastropods, crinoids and ostracods. Even farther offshore a pectinid facies was characterised by semi-planktonic forms such as *Dunbarella* and *Posidonia,* which were capable of surviving unfavourable bottom conditions either by periodic swimming or attachment to seaweed. Farthest offshore of all was a fauna characterised by nectonic goniatites. The Subcrenatum Marine Band, about 7 m thick in the Ashton Park area of Bristol, has a rich and varied fauna including several species of goniatites, but as it thins to the north and east the fauna becomes impoverished. The Aegiranum Marine Band is well developed throughout the district and shows a comparable thickness and faunal richness. The other marine bands mostly contain species of *Lingula* only.

Apart from those in the marine bands, the only other fossils that occur in any abundance are nonmarine bivalves. They are, however, less common than in South Wales where their use as zonal fossils was first developed. Other nonmarine fossils include ostracods, small branchiopod crustaceans such as *Leaia* and, more rarely, the cheliceratan *Eurypterus* and the remains of insects such as cockroaches and dragonflies.

THE COAL 'BASINS'

The Coal Measures are associated with a number of discrete structural areas, all but the Kingswood Anticline being informally described as coal 'basins' (Figure 12).

RADSTOCK SYNCLINE

The north-north-west-trending Radstock Syncline, lying to the south of the Farm-borough Fault Belt, is much broken by east–west thrusts or reverse faults (Figure 17), of which the well-known 'Radstock Slide' is an example (Figure 18). Dips are moderate towards the axis, which passes through Radstock and Timsbury. A small area of the Upper Coal Measures is exposed in the Farrington–Clutton–Timsbury area, and a considerable tract of the Lower and Middle Coal Measures occurs in the Nettlebridge Valley, but for the most part the productive measures are concealed by Mesozoic strata.

Several roughly north–south-trending faults, of which the '100 Fathom' or 'Clandown' Fault is the main one, are later than the east–west thrusts. Against the ridge of the Mendips at the south end of the syncline the inclination of the strata in-creases when traced eastwards from the old Moorewood Colliery, near Blacker's Hill, until, west of Barlake, the measures become vertical or overturned, a condi-tion that is maintained through Newbury, Vobster and Mells. The incompetent argillaceous Lower and Middle Coal Measures hereabouts display very complex contortions.

The time relationships of the Radstock Syncline strata were not properly under-stood until the discovery of the Aegiranum and Cambriense marine bands during the dismantling of the New Rock Colliery in 1968. This proved that the base of the Pennant Measures here is at a lower stratigraphical level than farther north (Figure 13). Only in the southern part of the Radstock Syncline have the coals of the Lower and Middle Coal Measures been worked, mainly at the Newbury and Vobster collieries in the south-east and in the New Rock and Moorewood pits to the south-west. Owing to the intense faulting and the almost complete abandon-ment of the workings, identification and correlation of the seams between the two areas is uncertain. In the Newbury and Vobster district some twelve seams vary-ing from 0.75 to 1.9 m in thickness were extensively worked. Some of the thick seams, such as the Dungy Drift and Coking Coals, appear to split and thin out westwards so that in the New Rock–Moorewood area only eight seams proved workable. In both regions the Perrink (Blackstone), Main Coal (Callows), Great Course and Garden Course seams have been extensively mined—the last two are in the Pennant Measures, above and below the Cambriense Marine Band respectively.

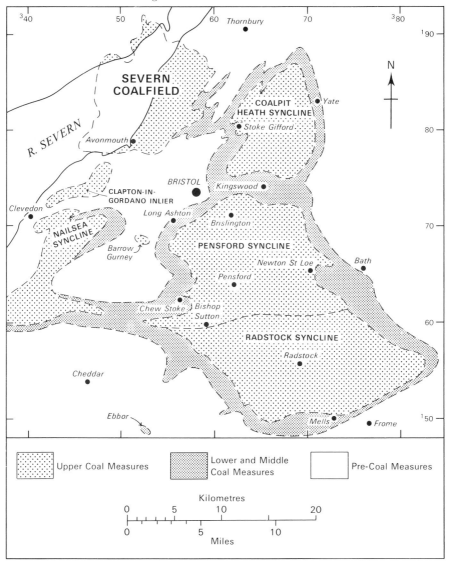

Figure 12   Distribution of Coal Measures (outcrop and subsurface) in Avon and Somerset, showing coal 'basins'.

Palaeontological knowledge of the Lower and Middle Coal Measures is very limited. The mudstones above the Perrink contain abundant large *Carbonicola* of *pseudorobusta* type, thus identifying the faunal belt of the same name (Table 3). Marine beds in contorted strata around the position of the Coking Coal, may represent the Vanderbeckei Marine Band.

The Lower and Middle Coal Measures are about 580 m in thickness, of which the top 150 m are placed in the Pennant Measures, which total about 1100 m in thickness. None of the coal seams above the Warkey Course in the Pennant Measures has been recognised in all parts of the syncline, although a few higher

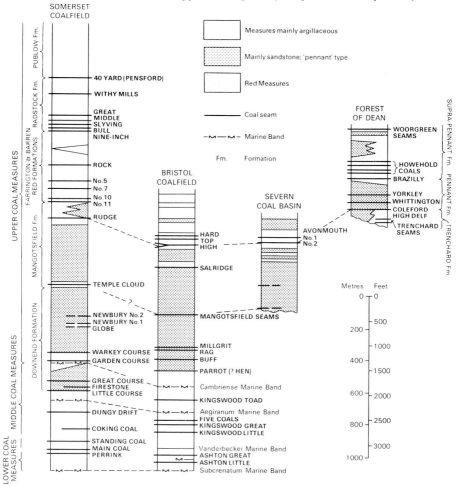

Figure 13 Comparative vertical sections of the Coal Measures in the Somerset, Bristol, Severn and Forest of Dean coalfields.

seams have been locally worked. These include the Newbury No. 1 seam (Figure 13), which contains *Anthraconauta phillipsii* in the roof. The remainder of the succession is not known in detail. The uppermost 400 m of the Pennant Measures crop out in the Clutton–Temple Cloud area, where they contain a few unworkable coals, the lower of which have very tentatively been equated with the Mangotsfield Seams of Bristol (Kellaway, 1970), thus allowing correlation with the Downend and Mangotsfield formations of the Bristol area.

The Supra-Pennant Measures, totalling some 900 m in thickness, occupy the central part of the Radstock syncline and are divided into the Farrington, Barren Red, Radstock and Publow formations whose boundaries are defined by coal seams. The coal seams in the Farrington Formation are thin, including the Rudge Vein at the base; only three of the five coals worked exceed 0.6 m in thickness. The Barren Red Formation, which has no workable coals, contains red beds, 90 m or so thick, in the middle of the formation in the southern and central areas; it thins northwards. The Radstock Formation includes six main seams, rarely exceeding 0.7 m thick, of which four were worked over most of the area. Little is known of the Publow Formation; up to about 130 m of grey measures with a number of very thin unworkable coals are present in the centre of the syncline. When traced northwards there is a tendency for all the seams in the Supra-Pennant Measures to split and deteriorate in quality, so that the coals in the Pensford area to the north, believed to be the equivalents of those at Radstock, are thinner and dirtier.

All the Supra-Pennant Measures contain typical Westphalian D floras. The Radstock Formation contains, in addition to an abundant flora, the bivalves originally used to define the *Anthraconaia prolifera* Biozone. However, owing to its erratic occurrence, it is now proposed that usage of the zone be discontinued (Ramsbottom et al., 1978); thus the whole of the Supra-Pennant Measures here belongs to the *tenuis* Biozone.

Pensford Syncline

This part of the Somerset Coalfield lies north of the Farmborough Fault Belt (p.72) and passes northwards into the Kingswood Anticline. Lower and Middle Coal Measures are known predominantly from scattered boreholes. The only outcrops of strata beneath the Supra-Pennant Measures are in the Newton St Loe area west of Bath where the Pennant Measures have a small outcrop and five seams of probable Middle and Upper Coal Measures age were worked from the Globe Pit in the 19th century. Elsewhere, only Upper Coal Measures rocks are exposed. On the northern edge of the syncline, the lowest Supra-Pennant Measures, presumed to be part of the Farrington Formation, are exposed in the Brislington area and were mined in the 18th and 19th centuries. On the west side of the syncline, four seams, also thought to belong to the Farrington Formation, were mined at Bishop Sutton until 1926.

The succession generally is imperfectly known except in the region around the Pensford and Bromley collieries. Here, above thick Pennant Measures two coal groups, the Bromley below and the Pensford above, are separated by barren measures including red beds; they are overlain by thick barren measures known as the Publow Formation, for which this is the type area. The thickness of the succession is appoximately equal to equivalent strata in the Radstock Syncline.

There are seven Bromley coals, of which three, 0.5 to 0.6 m thick, have been extensively worked. Out of eight coals in the Pensford coal group only two, both as thick as the Bromley seams, were widely worked. Exact correlation with the Radstock sequence is not possible because the coals used to define the formations there cannot be definitely recognised farther north, and faunal and floral differences add to the uncertainty. Using mainly lithological criteria, it is likely that the Bromley and Pensford formations roughly equate with the Farrington and Radstock formations, and the red beds between the Bromley and Pensford coal groups with the Barren Red Formation. Details of the Publow Formation were provided by the Hursley Hill Borehole, drilled by the National Coal Board between Pensford and Whitchurch (Kellaway 1970, pl.1). The formation, some 500 m thick, comprises mainly grey mudstone and siltstone with some mappable sandstones and occasional thin coal seams. A few of the coals were dug at outcrop on a very limited scale. Nonmarine bivalves of the *tenuis* Biozone are present at the base of the formation and a flora of Westphalian D age is present throughout.

Kingswood Anticline

A wide belt of Middle Coal Measures crops out in the core of an east–west-trending anticlinorium, long known as the Kingswood Anticline. Despite complex folding (Figure 19) and major faults, some 20 seams, ranging from 0.3 m to 2 m in thickness, have been extensively worked here.

The coals range from Lower into Upper Coal Measures. The lowest is the Ashton Group of coals, which includes the well-known Ashton Great Vein and Red Ash, and which has been extensively worked in the Ashton district. A recent section through the Lower Coal Measures was provided by the Ashton Park Borehole (Kellaway, 1967) which proved a previously unrecognised marine band with *Lingula* at about 12 m below the Ashton Great Vein, and stunted represen-tatives of the *extenuata* faunal belt in the roof of the underlying Little Seam. Although not proved in situ there is evidence for the Vanderbeckei Marine Band overlying the Red Ash Coal. The most widely worked coals in the overlying Mid-dle Coal Measures were the Kingswood Little or Two-Feet, the Kingswood Great or Bedminster Great, and the Lower Five Coals from below the Aegiranum Marine Band and the Kingswood Toad above it. Although the Cambriense Marine Band has not been located in the Kingswood Anticline, it is probable that the base of the mapped Pennant Measures approximates to the base of the Upper Coal Measures. The lowermost coals of these measures contain good coking coals, of which the Parrot, Buff, Millgrit and Rag have been worked extensively in the Oldland–Warmley district in the southern limb of the Kingswood Anticline.

Coalpit Heath Syncline

Lying to the north of the Kingswood Anticline this north–south-trending syncline extends as far north as Cromhall. Bedding dips are generally less than 40°. The syncline is divided into four parts by two intersecting faults. The north–south-trending Coalpit Heath Fault, slightly to the west of the fold hinge, has a downthrow of 100 m to the east; the Kidney Hill Fault lies at right angles to it and has a downthrow to the south.

The total thickness of the Lower and Middle Coal Measures is nearly 500 m. There are no workable coals in the presumed Lower Coal Measures but in the ex-posed north and north-eastern limb of the syncline the Middle Coal Measures have been worked on a limited scale in the Yate district. In marked contrast to the Kingswood Anticline, only two seams were mineable to any extent, i.e. the Yate Hard Vein and the Smith Coal, which correspond to coals in the Kingswood Great group.

Knowledge of the Lower and Middle Coal Measures and overlying strata in the southern and western part of the syncline was greatly extended after 1949 by a pro-gramme of nine deep exploratory boreholes drilled by the National Coal Board, which led to the opening of the Harry Stoke Drift Mine in the Kingswood Great group of seams. The boreholes enabled the accurate positioning of the main marine bands within the sequence.

The Upper Coal Measures outcrop is extensive and coals in the Supra-Pennant Measures were intensively worked and are virtually exhausted, whereas those below have been only selectively worked.

The Pennant Measures, which lie between the Cambriense Marine Band and the High Vein, vary in thickness from over 1000 m in the south to about 600 m in the northern part of the outcrop. The basal 120 to 180 m are markedly argillaceous, but, apart from some 40 m of mainly red beds underlying the High Vein at the top, the strata are dominantly composed of sandstone with relatively minor intervals of mudstone.

Using the Mangotsfield seams, which have been worked to a limited extent in the south, Kellaway (1970) divided the Pennant Measures into the Mangotsfield Formation above and the Downend Formation below (Figure 13). The sandstones of the former are uniform throughout the area. The north and north-eastward diminution in thickness of the succession, therefore, is thought to be due to the disappearance northwards of the main arenaceous division of the Downend Formation, which is represented only by a basal more argillaceous facies in the Rangeworthy area (Kellaway, 1970, fig.3). Coal workings are confined to the lower half of the Downend Formation adjacent to the Kingswood Anticline and the Mangotsfield seams. The lower coals, of which the Hen (0.9 m) is the most consistently developed, are apparently equivalent to the seams worked more extensively on the south side of the Kingswood Anticline. Elsewhere the coals are unworkable and red beds are commonly developed in this part of the sequence. There is no faunal evidence of age for the strata above the Mangotsfield seams, whose roof measures contain *Anthraconauta phillipsii* in the Mangotsfield area, which indicates the *phillipsii* Biozone.

The Supra-Pennant Measures have been correlated with the Farrington and Barren Red formations of the Somerset Coalfield. The Farrington Formation contains three workable coals, the lowest of which, the High Vein at Coalpit Heath Colliery, averages about 1.5 m in thickness with a thin parting. Southwards this seam splits and is represented at the Parkfield Colliery, south of the Kidney Hill Fault, by the Hollybush and Great Seams, separated by 0.4 m of shale. The shale increases to 15 m at the southern end of the area. The Farrington seams are succeeded by up to 275 m of barren measures, mainly red-brown but including an intercalation of grey measures in the upper part. Faunally, the roof measures of the High Vein are characterised by an abundance of the branchiopod crustacean *Leaia* in association with *Anthraconauta phillipsii* and *A. tenuis*. The roofs of the seams carry a rich Westphalian D flora.

Nailsea Syncline

Coal workings here were abandoned between 1880 and 1890 because of heavily watered measures and the inferior quality of the coal, and there is little information about the succession. About 270 m of shales and subordinate sandstones with 12 recorded coal seams are presumed to be Lower and Middle Coal Measures. Of the 12 seams only White's Top (1.1 m) and the Dog (0.9 m) were mined to any extent. A maximum thickness of about 330 m of sandstone, presumably Upper Coal Measures, rests on the Middle Coal Measures. Two seams are known here, but only one of these, Grace's Seam (0.9 m thick), was worked.

A short distance to the north of the main coalfield, Pennant Measures rest directly on Carboniferous Limestone with or without the intervention of thin Namurian quartzitic sandstone, thus testifying to strong sub-Namurian and sub-upper Westphalian unconformities.

Clapton-in-Gordano Inlier

Coal was worked on a very limited scale in this small and partly concealed inlier, but little is known of the structure. The succession is apparently entirely Pennant

Measures which overlie older Carboniferous and Devonian strata with marked unconformity.

Severn Coalfield

Apart from small inliers surrounded by Mesozoic rocks the coalfield is wholly concealed. Much disturbed Lower and Middle Coal Measures are exposed at Cattybrook, south-west of Almondsbury, and sandstones of the Pennant Measures crop out in reefs on the Welsh side of the River Severn and on the coast at Portishead and at Kings Weston. Much of the coalfield lies under the River Severn. Apart from the Cattybrook inlier, nothing is known of the lower part of the succession, though the Lower and Middle Coal Measures appear to be absent west of a line drawn between Olveston and Henbury, due to the unconformable overlap of the Upper Coal Measures. The lower subdivision of the Upper Coal Measures, 300 to 400 m thick, consists of massive sandstone (Pennant Measures) with minor shaly and coaly intercalations and, towards the top, red beds. An upper subdivision, proved by boreholes in the centre of the Avonmouth Syncline to exceed 160 m, consists of grey and some red measures which in the lowest part include two workable coals, the Avonmouth No. 1 and No. 2 seams.

A coal, 1.4 m thick, near the base of the Pennant Measures west of the River Severn, proved in boreholes in the Portskewett area, has roof measures with *Anthraconauta phillipsii*. Kellaway (1970) tentatively correlated this coal with the Mangotsfield seams of Bristol. The strata between Avonmouth No. 1 and No. 2 seams, and below the latter, contain abundant *Anthraconauta phillipsii, A. tenuis* and *Leaia*, and are correlated with the Hollybush and Great Veins of Parkfield in the southern part of the Coalpit Heath Syncline. The associated strata are therefore considered to represent the Farrington Formation, and possibly part of the Barren Red Formation within the *tenuis* Biozone.

Forest of Dean Coalfield

In contrast with the basins of the Somerset and Bristol coalfields, the Forest of Dean Coalfield, some 90 km^2 in area, is completely exposed. Two roughly north–south folds, the Main Syncline in the east and the shallow asymmetric Worcester Syncline in the west, are separated from one another by the Cannop Fault Belt. This NNW–SSE fault zone includes up to 25 faults, each with a throw no greater than 17 m. In the centre of the Main Syncline the strata are almost flat, but they become inclined on its eastern side in the Staple Edge Monocline.

The oldest Westphalian rocks occur in the north-eastern corner of the Main Syncline in the Mitcheldean area, where the upper part of the Drybrook Sandstone, lying unconformably beneath the Upper Coal Measures, contains miospores of Westphalian A (Langsettian) age. Over most of the area, however, the oldest Westphalian rocks comprise the Trenchard Formation, 15 to 120 m thick, at the base of the Coal Measures and resting uncomformably upon older formations (Figure 14). In the Coleford area the formation contains two coal seams which, farther south-east come together to form the 1.4 m-thick Trenchard Seam. Southwest of a line drawn roughly north-west–south-east through Coleford, the Trenchard Formation consists of mainly grey sandstones, but to the north-east it passes into barren red shales and mudstones. The lowest part of this formation may lie within the *phillipsii* Biozone, but there is no fossil evidence to prove this.

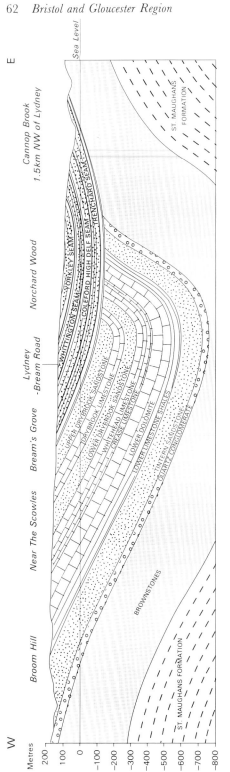

Figure 14   Section across the southern end of the Forest of Dean Coalfield.

The Pennant Formation ranges from 180 m in thickness in the north to 250 m in the south. At the base of this predominantly sandstone formation, the Coleford High Delf Seam, 1.15 to 1.5 m thick, has supplied 97 per cent of the total output of this coalfield since the Second World War. Two other seams in the Pennant Formation, the Whittington and the Yorkley, measure 0.8 to 0.9m in thickness, but are workable over only a limited area. Where the Coleford High Delf has a shale roof it contains a similar fauna to that associated with the Avonmouth No.1 and No. 2 seams and, like them, is referred to the *tenuis* Biozone. The flora associated with this and higher strata in the Pennant Formation indicates an early Westphalian D age.

The lower part of the Supra-Pennant Formation, about 90 m thick, is largely argillaceous and includes eight workable coals long known informally as the Household Coals. Many of the seams, however, are split into layers or 'leats' by soft mudstone partings and can usually be mined only where two or more leats have run together. The associated floras are of Westphalian D age. Above the highest of the Household Coals, known as the Crow or Dog, the upper part of the Supra-Pennant Formation attains some 240 to 250 m in thickness and includes an appreciable thickness of sandstone in its lower part. Near the top of the Supra-Pennant Formation, and preserved only in the centre of the basin, are the Lower and Upper Woorgreen seams (about 0.9 m). These beds contain a flora believed to be of basal Cantabrian age which, together with the Grovesend Beds of South Wales, makes them the youngest known coal measures in Britain (Ramsbottom et al., 1978).

The 'Horse' and 'Little Horse' in the Coleford High Delf of the Worcester Syncline are good examples of 'wants' or 'washouts', which extend for distances of 2.5 km and have an average width of 150 m and 50 m respectively. They occur where a coal has been locally replaced by sediments deposited from a former stream that crossed the coal-swamp and eroded away the peaty or coal deposits along its course. They are most abundant where the coal is overlain by sandy sediments.

OTHER COAL MEASURES OCCURRENCES

A small outcrop of Lower Coal Measures, including the Subcrenatum Marine Band, is present at Ebbor beneath the Ebbor Thrust, north-west of Wells. Other slices of Coal Measures are probably present beneath the Mesozoic cover within the Cheddar–Wells Thrust Belt on the south side of, and adjacent to the Mendips. In the absence of deep boreholes penetrating to the Palaeozoic basement, it is not known whether any appreciable areas of Coal Measures are present beneath the Mesozoic rocks of the Central Somerset Basin, though the geophysical evidence is thought to make this unlikely (Green and Welch, 1965).

Farther south and east, steeply dipping Coal Measures were proved in the Westbury Boring (Wiltshire), apparently on the southern side of the buried continuation of the Beacon Hill Pericline or an associated *en échelon* fold. The lithology and flora suggest that the strata are Lower or lower Middle Coal Measures.

Boreholes in the Moreton-in-Marsh to Burford area indicate that the western limit of the Oxfordshire Coalfield extends to the line of the Moreton 'Axis' (p.151). The coalfield succession comprises a thick coal-bearing grey sandstone sequence of Pennant Measure type in the lower part. There are two coal-bearing groups above, separated by barren and red measures, and a thick barren red measure sequence at the top. The total thickness is of the order of 1350 m. The fauna and flora show that the measures lie entirely in the *tenuis* Biozone of Westphalian D age. They unconformably overlie Upper Devonian but without any apparent angular discordance.

8   Intra-Palaeozoic earth movements

Much of the region lies within the Midlands Massif or Microcraton, which is bounded on the south by the Variscan Fold Belt. The massif was a stable region during the Palaeozoic Era, apparently undergoing only epeirogenic (i.e. largely vertical and extensional) movements up to the time of the Variscan orogeny. Evidence of Caledonian earth movement in the Lower Palaeozoic rocks is scanty, largely due to the restricted outcrop of the older rocks. The epeirogenic movements continued throughout the Devonian and early Carboniferous until the onset of compressional deformation during the Variscan orogeny. This last event had the most profound effect on the Palaeozoic rocks, and structures that developed during it and continued to influence sedimentation throughout the succeeding periods.

THE MALVERN 'AXIS'

The Malvern 'Line' or 'Axis' has long been regarded as one of the fundamental elements in the structure of the region, extending into it from the Welsh Borderland to the north. The 'Axis' is named after the narrow, elongate, north–south-trending horst of Precambrian igneous and metamorphic rocks that forms the Malvern Hills and which has been the site of uplift, including reverse faulting, at intervals since Precambrian times. The eastern boundary fault of the horst forms the western limit of the Mesozoic Worcester Basin in that area (p.150). At the southern end of the Malverns, the Precambrian rocks plunge out of sight beneath younger rocks, but are inferred from geophysical, mainly magnetic data, to extend south-south-eastwards beneath the Worcester Basin to near Tetbury, where they are estimated to lie at a depth of 3 km below the surface (Cave 1977, p.246).

The anticlinal axis at Huntley, to the east of Mitcheldean, the core of which is occupied by Silurian (Llandovery) rocks, and the *en échelon* Berkeley Fault and Coalpit Heath Syncline to the south, all trend in a north-north-westerly to north-north-easterly direction. They have either severally or together been referred to in the literature as manifestations of the Malvern 'Line' or 'Axis', but the name Malvern Fault Belt, recently coined by Kellaway and Hancock (1983), is more appropriate.

PRE-CARBONIFEROUS MOVEMENTS

Knowledge of early earth movements within the district is poor. Silurian rocks (Upper Llandovery) unconformably overlie the Cambrian (Tremadoc) in the Tortworth area; and, rather better known, is the interruption to sedimentation represented by the pre-Upper Devonian (Farlovian) unconformity. As a result of it, the thick Middle Devonian sequences of the areas to the south are absent from most of the region, while gentle folding accompanied by faulting and erosion has resulted in local, but appreciable overstep by the Upper Old Red Sandstone onto older formations. Movement was strongest in the Tortworth area, where the pre-Farlovian strata have undergone horst-type uplift aligned in a north–south direction that has resulted, following much subsequent erosion, in the Upper Old Red Sandstone directly overlying Wenlock strata in the central part of the uplift. Due to the cover of later rocks, only the western side of the structure, comprising the Berkeley Fault, is clearly seen; this has a pre-Farlovian downthrow to the west of some 600 m (Cave, 1977, fig. 22). Apart from this uplift, the early Palaeozoic rocks of the Tortworth inlier are generally more faulted and folded than the adjacent Carboniferous rocks, probably due to Caledonian movements, but the overwhelming effect of the later Variscan movements makes it hard to be sure. South of the Tortworth area, the Upper Old Red Sandstone in the eastern Mendips rests directly on Silurian (Wenlock) strata with low angular unconformity, and it has been suggested that the two areas lie on a north–south axis of uplift related to the Malvern Fault Belt.

INTRA-CARBONIFEROUS MOVEMENTS

During much of the Carboniferous period, the two sinking areas of thick continuous sedimentation represented by the South Wales Coalfield in the west and the Bristol and Somerset coalfields in the east were separated by an area in which the Carboniferous successions may show attenuation and nonsequence, and include important unconformities at the base of the Namurian, within the Westphalian and possibly also within the late Visean (p.47). Uplift and concomitant erosion appear to have been most active at the margins of the area, adjacent to the main coalfields, namely along the Malvern and Lower Severn 'axes' on the east and the Usk and Vale of Glamorgan 'axes' in the west (Figure 11). It has recently been suggested that these earth movements are related to an important dextral strike-slip fault, postulated to lie beneath the Severn estuary, analogous to the well-known Vale of Neath and Swansea Valley 'disturbances' farther west (Wilson et al., 1988).

The combined effect within the district of these movements is illustrated in Figure 11, which shows a diagrammatic reconstruction of the outcrops on which it is thought the higher Westphalian strata (probably Westphalian C = Bolsovian) were deposited. Folding, and hence angular unconformity, is confined to the belt of maximum uplift, and it is most strikingly shown in the Forest of Dean (Figure 14). It is probable that the late Westphalian measures formerly overstepped onto Lower Palaeozoic strata a short distance to the east of the present coalfield (Figure 14). The evidence for an early Westphalian age for part of the Drybrook Sandstone (p.61) indicates that the main intra-Carboniferous folding along the line of the Lower Severn–Malvern 'Axis' was mid-Westphalian in age. This episode of folding is regarded as the first pulse of the main Variscan orogenic movements. There is evidence to suggest that Coal Measures may attenuate eastwards along the eastern side of the Bristol Coalfield, and it has been suggested that a southwards continuation of the Malvern Fault Belt adjacent to the eastern margin of the coalfield may have been a positive area analogous to the Lower Severn 'Axis' (Kellaway and Welch, 1948; Kellaway and Hancock, 1983).

PERMO-CARBONIFEROUS MOVEMENTS

The main Variscan earth movements represent the most cataclysmic event to have affected the rocks of the region. The main movements took place between late Carboniferous (Stephanian) and earliest Permian times, shown by evidence in south-west England. In the present region the youngest deformed rocks are the Cantabrian Coal Measures of the Forest of Dean, and the oldest undeformed rocks are Permian strata proved at depth beneath the Mesozoic rocks of the Central Somerset Basin.

The region spans the Variscan Front, to the south of which the Devonian and Carboniferous rocks are folded and thrust-faulted along dominantly east – west lines. Eastwards from here seismic reflection evidence is now available to show a cross-section of the front at depth beneath the cover of Mesozoic strata (Figure 16). Between Belgium and south-west Wales the foreland area to the north of the front is characterised by a chain of late Carboniferous coal measures basins whose development is linked to that of the rising Variscan Fold Belt to the south. In the present region these comprise the Forest of Dean, Bristol – Somerset, Severn and Oxfordshire coalfields. South of the front, the Variscan Fold Belt within the region is divisible into two zones, a northern zone comprising the former area of Old Red Sandstone and Carboniferous Limestone shelf sedimentation, and a southern zone with thick marine Devonian and basinal Carboniferous shale ('Culm') sequences. The major anticlinal fold depicted in the southern zone is thought to be structurally analogous to the North Devon – Quantocks anticline (see p.147). The rocks of the Variscan Fold Belt are highly disturbed and commonly cleaved, whilst those of the foreland are much less disturbed.

The structural pattern within the region (Figure 15) indicates that the trend of the folds veers anticlockwise from an east – west trend within the Variscan Fold Belt in the south, to a nearly north-west – south-east trend in the north. In conformity with these trends, the steeper limbs of the folds change from north-facing in the south to south-west-facing in the north. These changes may be related to increasing distance from the main Variscan fold belt and show the increasing influence of inherited basement structures. Thus the north-easterly trends west of Bristol appear to reflect the influence of the Lower Severn Axis, while the northerly directed trends farther north appear to be related to the Malvern Fault Belt.

The base of the Carboniferous in the Pensford and Radstock areas is estimated to lie at from 2800 m below OD to a maximum of 3600 m, and in the deepest part of the Coalpit Heath Syncline the figure is about 2300 m. Although much of this depression is due to epeirogenic movements, mainly during the Carboniferous period, it cannot all be so explained and an important structural basin must have formed during an early phase of the Variscan movements. This was subsequently much modified by east – west-trending structures belonging to the later, main phase of the movements.

Cannington – Mendips

The largest known Variscan fault of the district lies to the north of Bridgwater, where Namurian strata are downthrown against the Middle Devonian Ilfracombe Beds on its north side, a stratigraphical interval estimated to represent at least 4 km of strata. The extensive Mesozoic cover unfortunately obscures the detailed relationships and doubt must remain as to the nature of the fault, but the most likely possibility is a high-angle reverse fault, possibly low-angle at depth and lying *en*

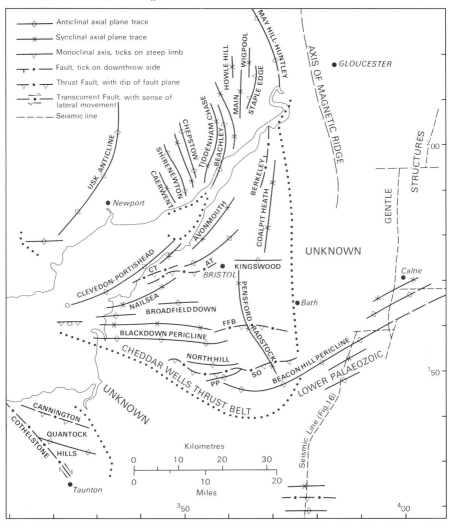

Figure 15   Major structural features in the Palaeozoic rocks of the region and of the adjoining area to the west as far as the Usk Anticline.

Abbreviations: AT   Avon Thrust, CT   Clevedon Thrust, FFB   Farmborough Fault Belt, PP   Pen Hill Pericline, SO   Southern Overthrust.

échelon with that recently recognised on seismic reflection evidence to the east (Figure 16; Chadwick et al., 1983). No further exposures are seen until the Mendips are reached, though the seismic data indicate the presence of large folds involving the Tremadoc-to-Carboniferous succession.

The Mendip ridge (or 'axis') constitutes an envelope of Old Red Sandstone and Carboniferous Limestone that wraps around the south and south-western edge of the Radstock Syncline. It consists of four *en échelon* periclines, the axes of which run approximately east–west. The folds are strongly asymmetrical, with steeper northern limbs that may be vertical or even overturned towards the north. Southerly dipping displacements parallel to the fold axis comprise thrusts or reverse faults associated with the steep limbs, and normal or lag faults associated with the gentler dipping southern limbs. Both these fault-types may end against or pass laterally into faults that cut across the fold axes at a high angle and which may have an appreciable component of lateral displacement. A slightly later phase of movement is represented by the Cheddar–Wells Thrust Belt which consists of several folds and thrust slices that were apparently piled up against the south-western side of the Mendip 'Axis'. In the Mendip area as a whole the limestones and sandstones involved in the structures behaved competently, giving rise to mainly concentric folding, although in local conditions of extreme stress, as in parts of the Cheddar–Wells Thrust Belt, the Lower Limestone Shale behaved in-competently and formed a plane of décollement above which the Carboniferous Limestone was folded independently of the underlying Devonian strata.

An analysis by Williams and Chapman (1986) of the Mendip–Bristol area, us-ing rigorous geometrical reconstruction techniques, has re-interpreted the struc-tural evidence in terms of 'thin skin tectonics' similar to those described from the Appalachians and elsewhere. It is postulated that the observed thrusts flatten at depth and join a gently (3°) dipping plane of décollement that underlies the whole area at a depth estimated to range from 3 to 5 km. The klippen of Carboniferous Limestone (see below) that lie in front of the Mendips are considered to be the remnants of 'first generation' thrusts that outcrop farther south but were folded as 'second generation' folds and thrusts formed in a 'piggy back' development.

East of the exposed coalfields

The eastern side of the Radstock Syncline is virtually unknown because mining never extended eastwards of the incrop of the Radstock Formation beneath the Mesozoic cover rocks. It was conjectured by Welch, in a classic paper (1933), that a mirror image of the Mendip ridge with *en échelon* periclines of Old Red Sandstone and Carboniferous Limestone might be present to the east of the syncline. Since that time, data from an aeromagnetic survey, boreholes and, most recently, a seismic reflection survey indicates that this is probably substantially correct (Figure 16), except that the individual folds are aligned in a north-east–south-west rather than an east–west direction. Large areas of Tremadocian rocks are now known to be present immediately beneath the Mesozoic cover east of the folded Devonian and Lower Carboniferous rocks that represent the eastern continuation of the Mendip Axis.

Radstock and Coalpit Heath synclines

The structure of the Radstock Syncline is markedly different from that of the Men-dips; the folding is relatively gentle, apart from the southern end, and the release of stress was predominantly by faulting. This difference is due to the strongly in-competent nature of the Lower and Middle Coal Measures which allowed the thick mass of competent sandstones comprising the Pennant Measures to fold in-dependently of the strata below.

This is most clearly seen at the southern end of the coalfield where the Lower and Middle Coal Measures are flanked on their south side by the steeply dipping Carboniferous Limestone, which comprises the northern limb of the Beacon Hill Pericline, and on their north side by a huge fold of regularly dipping Pennant Measures that are inverted over much of their length (Figure 17). The intervening Coal Measures are strongly contorted and squeezed. At Vobster and Luckington,

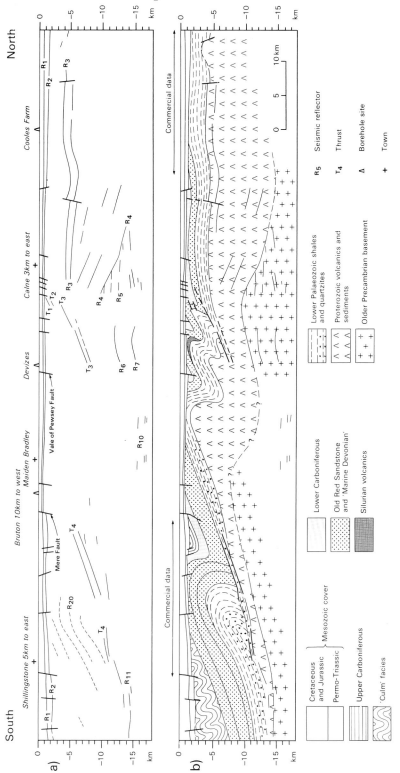

Figure 16 North–south sections showing a) principal reflectors identified from seismic reflection sections and b) geological interpretation of reflection events (after Chadwick, Kenolty and Whittaker, 1983).

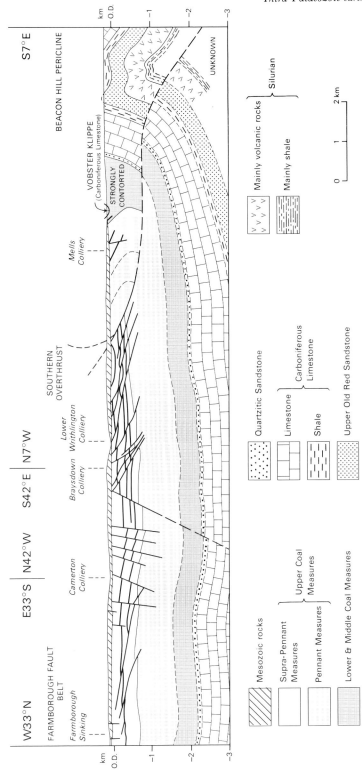

Figure 17 Horizontal section across the Radstock Coal Basin. The structure at depth remains unproved because no workings penetrate to the Pennant Measures north of the Southern Overthrust.

klippen of Carboniferous Limestone rest directly on these measures, which have been mined underneath. The mode of structural emplacement of these klippen, and apparently similar ones north of the Blackdown and North Hill periclines at Churchill and East Harptree respectively, has long been a matter for debate. One explanation has been given above; another (Figure 19) postulates the development of 'knee' folds in the incompetent Coal Measures that formed in front of the rising periclines and which, as folding proceeded, led to large-scale inversion, then stretching, and finally gravitational gliding. These processes were no doubt aided by the weight of the superincumbent flap of Pennant and higher measures in the upper part of the fold, which may itself have become detached from the crest of the fold before folding ceased. The front of the Pennant fold is terminated by a reverse fault, or thrust, known as the Southern Overthrust (or Great Southern Overthrust) which has an estimated upthrow to the north of at least 750 m. It has generally been considered, hitherto, that the overthrust died out within a fairly short distance to the south, but recent seismic evidence along the strike to the east indicates the likely persistence of thrust faults at some considerable depths (Figure 16). This supports the supposition that the Southern Overthrust is represented westwards, beyond the confines of the coalfield, as the Emborough Thrust, and then as the South-Western Overthrust in the Cheddar – Wells Thrust Belt even farther to the west.

Due to intensive mining in the past, the structure of the Upper Coal Measures is best known in the central parts of the Radstock Syncline, northwards from the Southern Overthrust. Faulting normal to the axis of the syncline is widespread and appears to be arranged in three belts. The southern half of the syncline is characterised by numerous low-angle thrust faults dipping in a southerly direction and of which the Radstock Slide (Figure 18) is the best known. A similar, though smaller group of southerly-dipping overlap faults, with an overall northwards throw of around 200 m to 250 m, is present at the northern limit of the syncline, and is known as the Farmborough Fault Belt. In the middle and deepest part, between Braysdown and Dunkerton, there is a swarm of east – west- to ESE – WSW-trending normal faults that dip to the north and rotate successive blocks downwards to the south, and which may possibly be related to underthrusting beneath. Cross-cutting these faults, there is a later series of north – south-trending normal faults subparallel to the axis of the syncline that apparently represents a final tensional phase of earth movement. The Clandown and Luckington faults are the most important of these; the former has a maximum downthrow to the west of 220 m.

The known Coal Measures of the western and northern parts of the Pensford Syncline are relatively far less disturbed than those of the Radstock Syncline but, farther north, the Kingswood Anticline (Figure 19) provides a striking contrast in which steep disharmonic folding is accompanied by thrust faulting, not only from the north and south but also from the east and west (Kellaway and Hancock, 1983, p.102). The structure appears to owe its prominence to the incompetence of the Lower and Middle Coal Measures, for the structure in the adjacent massive Pennant Measures on either side is much simpler. Along the main axis of the anticline, it appears that the earliest stress relief was upwards and, indeed, some of the folds are diapiric, i.e. they burst through the overlying strata.

Farther north the Coalpit Heath Syncline exhibits a relatively simple structure, with prominent north – south and east – west-trending tensional faults in its central parts. The syncline conforms to the regional structural pattern with its north – south elongation and appreciably steeper dips (up to 40°) on its eastern limb.

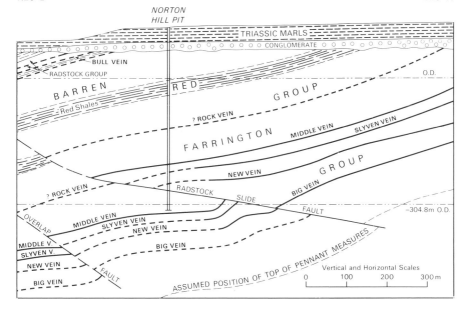

Figure 18 Overthrust faulting. The 'Radstock Slide' at Norton Hill Colliery, Avon.

West of Bristol

North of the Mendips and west of the main coalfields, the exposed Devonian–Carboniferous successions comprise mainly competent limestones and sandstones, the Lower and Middle Coal Measures having been largely overstepped by the Pennant Measures (pp.52–53); hence the folding is typically concentric, except in areas of most intense deformation. The belt of most complex structure, including important thrusting from the south and the south-east, stretches from Clevedon and Portishead, through the Avon Gorge, to King's Weston and Henbury, and thence to Thornbury. Here, both the Lower Limestone Shale and the Clifton Down Mudstone may locally behave incompetently, thereby leading to disharmonic folding above and below these formations. Within this belt, on the King's Weston–Henbury ridge, the strata are highly disturbed and extensively overturned to the north and north-west, while the intense folding of the Lower and Middle Coal Measures in the Cattybrook area, south-west of the Patchway Railway Tunnel, is associated with the north-east-trending Ridgeway Thrust Fault.

West of the River Severn, the progressive southwards anticlockwise swing of the fold axes reaches its greatest extent with the main folds in the Forest of Dean, which trend NNW–SSE. Faulting is normal and the folds are asymmetrical, with the steeper limbs facing to the west-south-west. The main Variscan folding is closely aligned, but not coincidental with that of the earlier intra-Carboniferous phase.

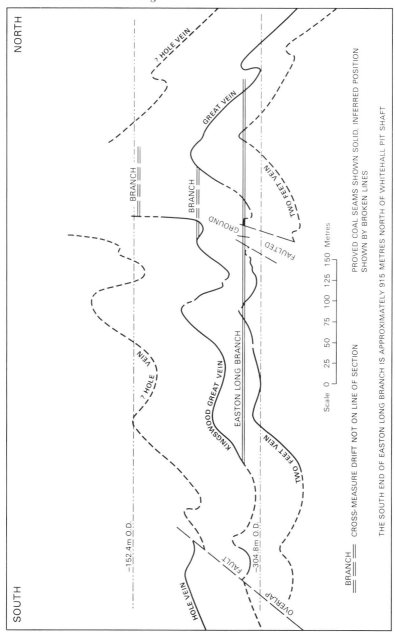

Figure 19   Section through part of the Middle Coal Measures at the western end of the Kingswood Anticline, Easton Colliery, Bristol (after E H Staples).

9 Permo-Triassic

Since the early days of geology the term 'New Red Sandstone' has been used to describe the dominantly red-bed continental rocks laid down in Britain during the Permian and Triassic periods. The survival of the term reflects the difficulty of subdividing these rocks. In the present region, gently dipping New Red Sandstone strata rest with marked angular unconformity upon a deeply eroded land surface of folded pre-Permian rocks. The outcrop distribution is shown in Figure 1.

The Permo-Triassic sandstones represent the thickest accumulation of any one sediment type to be deposited anywhere within the region in post-Carboniferous times, and the complete Permo-Triassic sequence is thicker than the sum of all the succeeding sediments in the region. The site of maximum sedimentation was within the Worcester and Central Somerset basins. These were actively subsiding grabens or half-grabens during early Permo-Triassic times, but during the middle and later part of the Triassic period sedimentation spread beyond their earlier confines. The thickest sequence is in the Worcester area, a short distance to the north of the present region, where some 2.5 km of sediments accumulated.

CLASSIFICATION

Table 4 shows the current and former nomenclature of the Triassic rocks of the district. The terms 'Keuper' and 'Bunter', which had both lithostratigraphical and chronostratigraphical connotations, have been replaced by new lithostratigraphical names (Warrington et al., 1980). Rhaetic, a term also bearing such connotations, has been abandoned and the name Penarth Group has been substituted. Rhaetian, a standard stage of the Triassic Period is not an equivalent of this. The other standard stages, which are based on ammonite stratigraphy in marine rocks elsewhere in Europe and Asia, are not generally applicable in the region, where fossils are scarce in the largely nonmarine sequence. Spores are potentially the best fossils for erecting a biostratigraphy despite their often poor state of preservation in the oxidised rocks.

Rocks of presumed Permian age occur locally at the base of the New Red Sandstone succession within the region, but the main part of it is Triassic in age. From the base, the Triassic succession is divided into three major divisions; the Sherwood Sandstone, Mercia Mudstone and Penarth groups. The base of the Sherwood Sandstone Group is marked by the first major widespread influx of pebbly debris, and is assumed to be approximately synchronous throughout the area. At outcrop, south of the region, the influx is marked by the Budleigh Salterton Pebble Beds, and in the Midlands by the 'Bunter Pebble Beds', the latter now given a variety of local names (Table 4). Between the outcrops in these areas, scattered boreholes in the south-west and north of the present region have encountered gravelly and pebbly beds at this stratigraphic level.

Table 4   Nomenclature of Triassic rocks

Stages	Groups	Formations		Members	Former names	
		South-West	S Midlands		South-West	S Midlands
	Lower Lias (basal part only)				Pre-planorbis beds ≡ Ostrea Beds	
					Watchet Beds	*not present*
Rhaetian	Penarth Group	Lilstock Formation		Langport Member	Langport Beds ≡ White Lias	
				Cotham Member	Cotham Beds	
		Westbury Formation			Westbury Beds	
– – – – – Norian	Mercia Mudstone Group	Blue Anchor Formation			Grey Marl & Tea Green Marl	Tea Green Marl
– – – – – Carnian – – – – – Ladinian – – – – –		Undifferentiated, mainly red mudstone		*see text*	Red Marls (Keuper Marl)*	
Anisian ? Scythian	Sherwood Sandstone Group	Otter Sandstone	Bromsgrove Sandstone		Upper Sandstones	Keuper Sandstone
			Wildmoor Sandstone			Upper Mottled (Bunter) Sandstone
		Budleigh Salterton Pebble Beds	Kidder-minster Formation		Pebble Beds & Conglom-erates	Bunter Pebble Beds

[P E R M I A N]

* Some usages included the Grey and Tea Green marls within the Keuper Marl.

The top of the Triassic is now taken immediately below the first appearance of the ammonite genus *Psiloceras*, which occurs slightly above the base of the Lower Lias.

CONDITIONS OF DEPOSITION

The Variscan orogeny created the floor of Palaeozoic rocks on which the Permian and Triassic sediments accumulated. The early stages in the process of erosion of this landmass cannot be reconstructed, but it is known that by early Triassic times the Coal Measures had, in places, been stripped off to expose Carboniferous

Plate 7 Unconformities at Woodhill Bay, near Portishead, Avon.
A. Between the Upper and Lower Old Red Sandstone (Devonian) with associated
cornerstone calcrete (A10737).
B. Between the Dolomitic Conglomerate (Triassic) and the Lower Old Red Sandstone
(A10732).

Limestone. Thus, when traced from the east Devon coast towards Williton and Puriton, the basal Triassic conglomerates are found to contain an increasing quantity of pebbles of Carboniferous Limestone. In early Triassic times the Mendips and the Bristol Coalfield formed a hilly tract north of the Central Somerset Basin, with the west Somerset and Devon highlands on the west. In the north, another basin lay between the Malvern Fault Belt and the Vale of Moreton 'Axis'. The detritus from the eroded highlands, either borne by the wind or washed down by torrents after sudden storms, accumulated in the deeper parts of both these basins.

By late Triassic times a considerable thickness of sediment, comprising sandstone overlain by silty mudstone, had accumulated in this way. Thin beds of grey-green calcareous sandstone and silty sandstone ('skerries') in the latter indicate local, but often widespread, fluvatile and/or lacustrine incursions. At certain periods the concentration of brines by evaporation of saline lakes gave rise to deposits of gypsum and rock salt (halite), collectively known as 'evaporites'. At the same time, thick screes of angular and rounded rock debris accumulated on the mountain slopes and as outwash fans in front of the mountains, which flanked the basins. These deposits in the Bristol and Mendip areas are known as the Dolomitic Conglomerate (Plate 7B). Within the areas of high relief, such as the Mendips, the debris filled narrow wadis, 100 m or more in depth, comparable in dimensions to present day gorges such as Burrington Combe or Ebbor Gorge. Individual boulders, many tons in weight, are present in the screes banked against steep cliffs of Carboniferous Limestone, and can, for instance, be seen in the Bridge Valley Road section in the Avon Gorge and along the coast between Portishead and Clevedon.

The Dolomitic Conglomerate, which has an extensive distribution, passes laterally into rocks varying from early to late Triassic in age and is therefore regarded as a diachronous marginal facies. It was found resting on Carboniferous rocks in the western part of the Severn Tunnel, and is well developed in the Portskewett – Caldicot area. South-west of Chepstow it attains a considerable thickness and forms the English Stones, Ladybench and other reefs and benches in the Severn Estuary, where it rests upon Coal Measures sandstone. Flanking the Carboniferous Limestone over much of Gwent and parts of Broadfield Down and the Mendips, however, the marginal facies consists in places of hard pink and yellow impure dolomite that apparently represents highly altered, much comminuted primary carbonate debris.

Where breccias derived from the larger uplands passed outwards onto the more deeply eroded surface of the Coal Measures, huge spreads of Dolomitic Conglomerate were deposited. For example, a 13 km-wide deposit extends northwards across the relatively flat surface of the Coal Measures from the foot of the Mendips.

By late Triassic times the accumulation of rock debris on the slopes of the areas of high relief, and the deposition of sand and silty mud in the interior basins, had created a great plain and substantially reduced the surface relief. The closing phase of deposition of the red mudstones was marked by an increasing number of green beds, which may indicate a gradual amelioration of the climate as humid periods became longer and more frequent. There is good evidence that the sculpting of the pre-Permian land surface was due to desert erosion. Where the red mudstones and Dolomitic Conglomerate have been eroded away, the northern face of the Mendips rises at the present day as a great reddened wall of steeply dipping, patchily dolomitised limestone. Similar steep faces are found around Broadfield Down and in other denuded limestone hills. They are believed to be close to their Permian form. In the absence of a rainfall sufficiently persistent to maintain streams on the limestone areas or to remove the limestone in solution, stream erosion would tend to become concentrated on the softer and less pervious Coal Measures. In periods of long drought, wind erosion on the plain would also tend to maintain rather than reduce the slope of the limestone cliffs.

Well-rounded grains of quartz, of probable aeolian origin, and fragments of Carboniferous Limestone occur in the sandstone beds in the red mudstones. Together with extensive reddening of the sub-Permian rocks, they indicate desert conditions. Where the Dolomitic Conglomerate is absent and the red mudstones overlie the Coal Measures, a vivid red sandstone containing hematite (known as red ochre) commonly occurs at the base of the Mercia Mudstone Group. The hematite is believed to be derived from the oxidation of pyrite in the Coal Measures shales.

The fauna and flora of the New Red Sandstone indicate a terrestrial depositional environment. Few fossils are found in the red rocks. The bones of terrestrial reptiles, principally from fissure deposits, include small lepidosaurs and archosaurs, and include the sphenodontids *Clevosaurus hudsoni* and *Kuehneosaurus latus*. The deposits filling fissures in karst areas and palaeo-caves in Carboniferous Limestone are of uncertain age, but palynomorphs at one locality suggest a late Triassic age.

Fossils, including the crustacean *Euestheria*, the teeth of sharks *(Hybodus)* and the lung fish *Ceratodus*, and also trace fossils and plant remains including miospores, have been reported from the Arden Sandstone, the most widespread of the 'skerries' in the mudstone succession. A record of a unique tooth *(Hypsiprymnopsis rhaeticus)* from the late Triassic Blue Anchor Formation near Watchet, in west Somerset, was regarded by Boyd Dawkins (1864) as of mammalian origin.

Finally, in the late Triassic, the surface was inundated by the advancing waters of the Rhaetian sea, a widespread and important event usually called the Rhaetian Transgression. This heralded a period of marine sedimentation that was to last throughout most of the Mesozoic Era. The marine Rhaetian fauna, which appears in greatest numbers and variety in the black shale of the Westbury Formation, consists largely of thin-shelled bivalves, associated with echinoids, ophiuroids and gastropods. The succeeding Cotham Member of the Lilstock Formation marks the establishment of a shallow sea with a calcareous-mud floor or, locally, a lagoon only infrequently entered by marine organisms. When the water bodies dried out, mud-cracked layers were formed, while remaining pools of fresh or brackish water were colonised by algae, liverworts, crustacea and ostracods. Incursions by the sea led to the formation of thin beds with marine shells.

This period of intermittent marine incursions gave way to permanent marine conditions, represented by the Langport Member of the Lilstock Formation, and the Lower Lias.

PERMO-TRIASSIC SANDSTONES

Although the sandstones forming the lower part of the New Red Sandstone succession have been recognised at depth in scattered boreholes over the eastern and southern parts of the district, their outcrop is limited to a small area on the western faulted margin of the Worcester Basin at Huntley. Elsewhere, they are overlapped at the basin margins by the overlying Mercia Mudstone.

Within the region, seismic reflection and borehole evidence indicate that beneath the Gloucester – Vale of Evesham area, where the succession is thickest, the lowest strata comprise the Bridgnorth Sandstone, of possible Permian age. Overlying this is the Kidderminster Formation, of earliest Triassic age. The combined thickness of these two formations varies from about 400 m to 700 m, of which the conglomerates of the Kidderminster Formation account for up to about one-third. At outcrop beyond the district to the north and north-west, the Bridgnorth Sandstone comprises uniform, bright red, strongly cross-bedded dune sandstone. The overlying pebble beds are fluvial in origin and reflect the onset of markedly wetter conditions, consequent on major geographical changes beyond the margins of the region. The succeeding sandstones attain a thickness of some 1000 m, but the available evidence within the region does not permit recognition of the Wildmoor and Bromsgrove sandstone formations, which are distinguished farther north in the Midlands. To the north, the Wildmoor Sandstone is a relatively uniform red sandstone with few pebble layers and may have been partly aeolian in origin. The Bromsgrove Sandstone includes beds of pebbly sandstone and siltstone typically arranged in fining-upward fluvial cycles, and is dull reddish brown in colour, changing to whitish and yellow-brown hues in the upper part of the sequence; the formation was well represented in the Stowell Park Borehole near Northleach.

On the east side of the Worcester Basin an attenuated representative of the Bromsgrove Sandstone overlaps onto the London Platform and is itself overlapped by deposits of the overlying Mercia Mudstone Group farther east. On the west side of the basin, north of the River Severn at Sharpness, the Triassic is represented at outcrop by the Mercia Mudstone Group, which is strongly downfaulted against the Palaeozoic rocks. South of this the Mercia Mudstone Group margin is unfaulted and the limits of the main sandstone sequence beneath are unknown.

The basal sandstone sequence is absent in the Bristol–Mendip district, where the later Triassic rocks have overstepped directly onto the Coal Measures and older rocks. The sandstones reappear to the south within the Central Somerset Basin, which forms part of the south-west England 'province'. Permo-Triassic sequences there, with the exception of the Budleigh Salterton Pebble Beds, which are considered by many to equate with the various 'Bunter' pebble bed horizons of the Midlands, cannot be clearly related to those of the Midlands. The thickest sequence so far proved in the southern area is in a borehole at Puriton, north of Bridgwater, where the lowest strata, not bottomed, comprised 177 m of uniform, brick-red, fine-grained sandstone with green spots and blotches and some interbedded siltstone. These are overlain by 65 m of red and grey sandstones with interbedded pebble and siltstone layers, and a well-marked basal conglomerate about 4 m in thickness. Some pebbles of Carboniferous limestone occur in this conglomerate, which has been equated with the Budleigh Salterton Pebble Beds; thus, by definition, the underlying beds are Permian in age. The nearest outcrops of corresponding rocks are outside the region on the west side of the Quantock Hills, where a somewhat similar succession, totalling 230 m in thickness and including a conspicuous representative of the pebble beds up to 30 m thick, is faulted against the Palaeozoic basement.

MERCIA MUDSTONE GROUP

The Mercia Mudstone Group has long been known as 'Red Marl', a term often used synonymously with 'Keuper Marl'. The rocks consist largely of red dolomitic siltstone and mudstone with a starchy texture and a feebly conchoidal fracture.

Only slightly calcareous, these rocks do not warrant the use of the term marl. The red mudstones overlap the Sherwood Sandstone Group onto Carboniferous or older rocks round the sides of the Worcester and Somerset basins, and extend over much of the region either at outcrop or beneath a cover of later rocks. The greatest thicknesses, 450 to 550 m, occur in the central parts of the basins. In the western half of the region the Mercia Mudstone Group is variable in thickness because of the irregularity of the pre-Triassic land surface, which it masks. In the Bristol–Mendip area, it is locally absent where the overlying Penarth Group oversteps onto the remnants of the ancient hills (Figure 21).

In the Worcester Basin and areas to the north, up to 10 m of alternating mudstone and sandstone beds occur between the Sherwood Sandstone and Mercia Mudstone groups. These rocks were named 'Waterstones' from the fancied resemblance of their mica- spangled bedding surfaces to the shading of watered silk, and were formerly classified as the uppermost formation of the 'Keuper Sandstone'. Current usage incorporates them within the Mercia Mudstone Group and the name "Waterstones" has been abandoned. Apart from the highly micaceous layers, the mud cracks and small-scale, often contorted cross-bedding are features characteristic of these transitional beds.

The red mudstones commonly have small patches, streaks and occasional bands of green or grey-green, the colour differences being ascribed to the oxidation state of iron in the constituent minerals. Occasional beds of sandstone and hard siltstone ('skerries') form topographic features in the otherwise flat landscape. The most notable arenaceous development is the Arden Sandstone Member which occupies a position about one-third down the Mercia Mudstone Group succession throughout most of the Worcester Basin. It is present at outcrop in the district between Newnham and Tewkesbury, where in former times it constituted a local source of building stone. The thickness varies from one to as much as 7 m. The lithology comprises mainly white to fawn, calcareous sandstone and grey, green, red and purple mudstone, which occur in every combination from thick beds to fine interlaminations. Mudcracks and small-scale cross- bedding are common. The depositional environment may have been deltaic or estuarine, with sands deposited in distributaries that traversed broad mudflats. In the Bristol–Mendip area similar sandstones are present in the uppermost 8 to 25 m of the succession. The Stoke Park Rock Bed of the Bristol area and the Butcombe Sandstone north of the Mendips are the most widespread. In the Vale of Taunton, the North Curry Sandstone may be at approximately the same stratigraphical level. Palynological evidence indicates a Carnian age for these beds.

In the western half of the region the red mudstones pass laterally into a marginal facies that results from erosion of the older rocks (see above). The most widespread facies, known as the Dolomitic Conglomerate, was mainly derived from the Carboniferous Limestone but where there are very extensive outcrops of Coal Measures, as in the Bristol–Pensford area, the marginal facies may comprise soft red and fawn calcareous sandstones. In the Bristol area, where they are named the Redcliffe Sandstone, these may locally exceed 50 m in thickness. The red sandstones gave rise to the name Redcliffe, a district of Bristol, where they form river cliffs along the Avon and are well exposed on the south side of the New Cut between Bathurst Basin and Ashton Gate.

Primary and secondary evaporite deposits are widespread in the Mercia Mudstone Group. The largest single deposit is the Somerset Halite, which has a widespread occurrence within the Central Somerset Basin and may extend south-eastwards into the Wessex Basin in Dorset and westwards under the Bristol Channel.

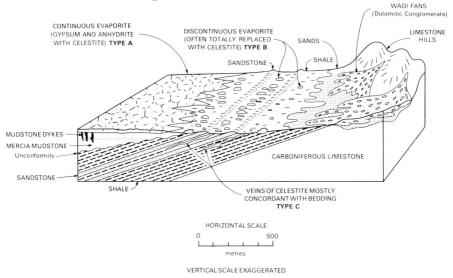

Figure 20   Diagram illustrating modes of celestite occurrences north of Bristol (after Nickless, Booth and Moseley, 1976).

In the Burton Row Borehole, halite or rock salt (sodium chloride) was found in four main beds in the middle of the Mercia Mudstone Group, over a vertical thickness of about 95 m of strata, with veins extending a further 30 m below. The thicknesses in the Puriton Borehole were rather less. Clay 'pseudomorphs' after rock salt occur at a number of places and horizons in the Mercia Mudstone Group.

Apart possibly from dolomite, which occurs ubiquitously as minute interstitial grains and not uncommonly as a secondary cement, the most widespread evaporite mineral in the red mudstone is calcium sulphate, either in the hydrated form, gypsum, at outcrop and shallow depths, or in the anhydrous form, anhydrite, at greater depths. It occurs in nodules, nodular bands and veins, and can be seen at Aust Cliff (Plate 8A), for example. The nodules are commonly about 30 cm across, but range from pinhead disseminations to large masses a metre or more across and weighing several tons. They are thought to have been formed by accretionary processes, just below ground surface, in a mudflat or inland sabkha-type environment under arid conditions with a high rate of evaporation. Both anhydrite and gypsum can form as primary minerals in this type of environment. The veins are due to secondary redistribution of calcium sulphate and re-precipitation in acicular or needle-like crystals normal to the sides of cracks and joints, both in the adjacent and underlying strata. The nodules are scattered throughout thicknesses of tens of metres of strata. Nodular beds, usually less than 30 cm thick, are occasionally present and apparently result from the coalescence of adjacent nodules. No thick, commercially mineable beds have been reported. Deep borehole evidence in the Worcester Basin suggests that gypsum/anhydrite nodules tend to be concentrated in two main horizons, one directly above and the other below the Arden Sandstone. In the Burton Row Borehole, in the Central Somerset Basin, although anhydrite nodules are particularly abundant in a 40 m-thick horizon directly above the Somerset Halite, nodules are scattered throughout the sequence.

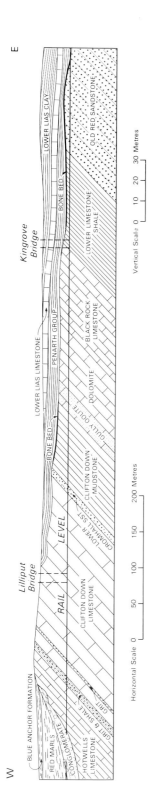

The Bristol area has long been noted for the occurrence of celestite (strontium sulphate) which has been exploited commercially at numerous localities during the last 100 years. The main stratigraphical occurrence is within or immediately adjacent to the Stoke Park Rock Bed (see above) and the celestite-rich unit is named the Severnside Evaporite Bed (Nickless et al., 1976). The thickness varies from 0.5 to 2 m. The outcrop has been traced over a wide area extending from Bitton in the south to the Thornbury and Charfield areas in the north. The primary mineral is thought to have been gypsum and/or anhydrite, and all stages of its replacement by celestite can be seen. The source of the strontium is unknown, although proximity to the Palaeozoic basement appears to be important. The various modes of occurrence are illustrated in Figure 20; in the first two the celestite occurs as nodules, fine-grained disseminations and veins, either within the Severnside Evaporite Bed itself (type A), or on the contemporaneously exposed land surface of Palaeozoic rocks (type B). Type C comprises veins, sheets and vuggy infillings apparently secondarily redistributed in the underlying strata, which may be either the Mercia Mudstone Group or the Palaeozoic basement.

In the Bristol–Mendip area, the strata adjacent to the sub-Triassic unconformity commonly contain masses of coarsely crystalline calcite, sometimes with specks of galena or sphalerite or, more rarely, bright green copper secondary minerals. Geodes or hollow nodules ('potato stones') lined with red, yellow and amethystine quartz crystals have achieved fame in the past under the exciting name of 'Bristol Diamonds'. Celestite, barytocelestite, barite and gypsum are commonly associated with these various mineral occurrences, both as primary constituents and as secondary replacements.

Figure 21 Section of railway cutting south of Chipping Sodbury, Avon.

Plate 8   Mercia Mudstone Group and Penarth Group successions (late Triassic)
A. Aust Cliff, Avon (A10669).
B. Blue Anchor, north Somerset (A11715).

Blue Anchor Formation

In west Somerset a series of magnificent coastal sections between Blue Anchor (Plate 8B) and Lilstock displays the Triassic succession from the upper part of the thick red Mercia Mudstone deposits to the basal Lias. The junction between the predominately red rocks below and the superincumbent grey and green strata of the Blue Anchor Formation is transitional, with the proportion of red beds diminishing upwards over some 50 to 60 m. The base of the Blue Anchor Formation is taken somewhat arbitrarily above the highest prominent bed of red mudstone, where the transition to green-grey coloration is complete. In the Central Somerset Basin, the Blue Anchor Formation is 20 to nearly 40 m thick and comprises alternating dark grey mudstones, some of which are shaly, and greenish grey or buffish grey silty mudstones and siltstones. The latter may be dolomitised and weather out as strong ribs in the coast sections. Finely laminated beds displaying burrows are present in places. Gypsum often occurs as nodules and veins, which, in the coast sections, are best developed at two main levels in the middle of the sequence.

Traditionally the formation has been divided into Tea Green Marl below and Grey Marl above; the latter has been distinguished from the former by the presence in it of dark grey mudstone bands. On the coast, the Tea Green Marl, as so defined, is rarely more than 5 m thick. Inland it is not usually possible to separate the two. The uppermost few metres of the Grey Marl were formerly separated as the Sully Beds, because they locally contain elements of the same fauna as the overlying Westbury Formation. However, the 'Sully Beds' are lithologically identical to the remainder of the Grey Marl and markedly different from the Westbury Formation, and the term has been abandoned.

Within a few kilometres south of the Mendips, along the northern edge of the Central Somerset Basin, the Blue Anchor Formation thins rapidly and on the Mendips it is mostly overlapped by the overlying Penarth Group. North of the Mendips the formation is 12 to 13 m thick in the Uphill–Locking area, where it includes some dark grey mudstones of 'Grey Marl' type in its upper third, but the transition to the red mudstone below is sharp. In this area, and in the marginal areas south of the Mendips, local occurrences of small masses of coarsely crystalline calcite and/or quartz and celestite, at about the base and the middle of the formation, may represent altered evaporite deposits.

Elsewhere in the region the formation is rarely more than 2 to 5 m thick. In the Worcester Basin it measures no more than 9 m and the dark grey mudstone facies is absent. The boundary with the underlying red mudstones is sharp and may represent a nonsequence. Beds of finely interlaminated mudstone, siltstone and fine-grained sand are usually present in some parts of the sequence but evaporite minerals are rare.

PENARTH GROUP

The Penarth Group is named after the coast sections at Penarth, South Glamorgan, and is virtually synonymous in a lithological sense with the former 'Rhaetic'. The group contains the Westbury and Lilstock formations, comprising thin, very distinctive rock sequences uniformly developed throughout much of Britain.

Westbury Formation

The Westbury Formation takes its name from Garden Cliff, Westbury-on-Severn. It is better developed in the Central Somerset Basin where it reaches a thickness of nearly 14 m. Here, the Blue Anchor Formation is succeeded by dark grey to nearly black shale with thin beds of limestone and sandstone. The junction is usually a nonsequence. The colour change from grey and green to black is striking, and the presence of fragments or pebbles of the underlying rocks in the base of the Westbury Formation shows that the former suffered erosion prior to the deposition of the black shales.

Thin-shelled bivalves are abundant in some beds of the black shales and include *Rhaetavicula contorta, Eotrapezium concentricum, Chlamys valoniensis* and *Tutcheria cloacina*; small gastropods, fish remains and other fossils also occur. Thin limestones full of *Chlamys valoniensis, Placunopsis alpina* and other forms can be traced locally, and form useful but limited aids to correlation. In this respect, one of the most interesting beds is the Ceratodus Bone Bed, a conglomeratic sandy limestone a few centimetres thick, which passes locally into ginger-coloured sandstone. The rock is packed with vertebrate remains, mainly the scales, teeth and spines of *Acrodus, Hybodus, Gyrolepis* and other fishes, together with the bones and teeth of marine saurians such as *Ichthyosaurus* and *Plesiosaurus*. Locally, the most characteristic remains are the large palatal teeth of Ceratodus, a relative of the modern lung-fish. Except in the Bristol–Mendip area, where the lowest beds have been overlapped against the basement, this bone bed occurs at the base of the Westbury Formation. In the Central Somerset Basin, where the formation is thickest, up to four or five thin bone beds, separated by black shales, are developed in the lower part of the sequence. In the Wedmore area, the lowest part of the formation includes a hard, shell-fragmental limestone, the Wedmore Stone, 0.8 to 1.4 m in thickness, that was formally much used as a local building stone.

In the Mendip area and the country to the north, the erosion that preceded the deposition of the Westbury Formation was more severe than in the south, and the widespread submergence that followed it led to the deposition of fine, dark grey muds on an eroded surface composed of a wide variety of formations (Figure 22). Fragments of green and red Mercia Mudstone Group sediments, and insoluble residues such as rolled phosphatic pellets and quartz pebbles are abundant in the basal bone bed or, where this is missing, may be seen lying on the eroded surface beneath the black shales.

There is no extensive littoral deposit of the type found in Glamorgan, but at Butcombe, north of the Mendips, the rocks pass into a conglomerate packed with well-rounded pebbles of Carboniferous Limestone. A Mendip shoreline may have contributed to this deposit, and the occurrence of terrestrial vertebrates, including reptiles and early mammals in late Triassic deposits infilling fissures in the Carboniferous Limestone of the Mendips area proves the existence of one or more island land areas at this time.

Lilstock Formation

The Lilstock Formation comprises a lower argillaceous unit, the Cotham Beds (now Cotham Member) and an upper dominantly limestone unit, the Langport Beds (now Langport Member); the latter is synonymous with the White Lias *sensu* William Smith in the Bristol district.

A nonsequence marks the junction of the Westbury Formation and the overlying Cotham Member. The latter consists of soft, greenish grey, silty, calcite mudstones at the top of which, in the Bristol district, lies the well-known Cotham Marble. The total thickness of the member is rarely more than 2 m. The location of the Cotham Member, and to some extent of the Westbury Formation also is, however, of importance to the engineer since these rocks have low bearing strength, particularly when weathered.

The fauna of the Cotham Member is generally impoverished, and in west and central Somerset very few fossils are to be found, except at the base. North of the Mendips it is slightly less barren and a thin bed of banded calcite mudstone in the lower part of the member has both plant- and shell-bearing layers. This, the Naiadita Bed, yields *Naiadita lanceolata*, a fresh- or brackish-water liverwort (Bryophyta) and its associated spores, together with algae, insect larvae and occasional specimens of the crustacean *Euestheria minuta*. The occurrences of *Euestheria*, of plants and of marine shells, including *Chlamys valoniensis* and other bivalves, are usually in separate distinct beds. Mud-cracks and, more rarely, ripple-marks and worm-tracks, also occur in the Naiadita Bed.

The Cotham Marble takes its name from the type locality at Cotham, Bristol. It is a hard, splintery calcite-mudstone with an irregular mammillated top and a smooth flat base. In Gloucestershire it passes locally into a hard, fissile limestone with *Meleagrinella fallax*, but elsewhere animal fossils, apart from fish scales, are rare. The bed is seldom more than 0.2 m thick and is notable for the tree-like markings with rounded cloud-like forms above, which are seen when the rock is broken in a vertical plane; hence the popular name 'Landscape Marble'. In the past it was used as an ornamental stone, both cut and polished for indoor use and, undressed, for outside walls and ornamental rockeries. It is now generally agreed that the 'Landscape' is an association of algal growths occurring in convex masses. Other types known as 'False' or 'Crazy' Cotham Marble were formed by the penecontemporaneous breaking up of partly consolidated calcite mud; these contain shell detritus including bivalves and echinoderm fragments, and may occupy channels between the algal 'buns' as lag deposits.

The Langport Member, or the White Lias as it has long been known in this district, consists of pale grey and cream limestone with mudstone partings. The limestones tend to be irregular and rubbly in the lower part of the succession and finer grained, even bedded and very hard in the upper part. The upper beds have been widely quarried in the Somerset Coalfield for use as a building stone. The top bed, locally called the 'Sun Bed', may show signs of emergence in the Somerset area, where it is typically penetrated from above by U-shaped tubes. The member has a restricted fauna of bivalves, gastropods and locally abundant ostracods, mainly in the lower part; echinoid fragments and simple corals occur sporadically. Some bivalves, such as *Modiolus* and *Protocardia*, are common in the underlying beds, but a few genera including *Astarte, Plagiostoma* and *Pleuromya* make their earliest regional appearance in the White Lias. Most of the genera and species persist into the Lower Lias. Small fronds and leaves, including those of the bennettitalean plant *Otozamites obtusus* and miospores of pteridophytes and gymnosperms, represent the land flora. The fauna indicates that the sea was very shallow, with the sea bed being intermittently exposed.

The Langport Member is thin or absent north and west of a line joining Stratford-on-Avon to Bristol thence running southwards to the Mendips and turning westwards along the northern margin of the Central Somerset Basin. At or near outcrop, the thickness varies from just under a metre to rather more than 5 m, but scattered borehole evidence suggests that it thickens eastwards and south-eastwards down dip to 8 m or more within a gulf extending northwards beneath the Cotswolds and southwards to the Wessex Basin. The thickness variations are due partly to depositional causes and partly to the effects of pre-Liassic and intra-Liassic erosion in different parts of the district. The remaining part of the Triassic System, which comprises the lowest few metres (at the most) of the Lias, is considered in the next chapter.

10 Lower Jurassic

The marine transgression at the end of Triassic times, during which the Penarth Group was deposited, was followed by the establishment of open-sea conditions, first indicated by the early Liassic strata but which continued for much of the Mesozoic era. Liassic rocks, comprising the whole of the Lower Jurassic, crop out as a wide north-north-easterly trending belt, much of it low-lying heavy clay land, that traverses the region. The succession continues eastwards beneath a cover of younger rocks. Sedimentation in the Central Somerset Basin and beneath the Cotswolds was more or less continuous. Between these areas, in the Bristol – Mendip area, the succession is thinner and sedimentation was interrupted. The eastern edge of the main Cotswold Basin (a continuation southwards of the Worcester Basin) is formed by the Moreton 'Axis', which also defines the edge of the gently and intermittently subsiding London Platform, a low-lying landmass in Jurassic times, which extended eastwards beyond the Oxford area as far as Belgium.

CLASSIFICATION

The base of the Jurassic System is defined by the first appearance of ammonites of the genus *Psiloceras* and, therefore, the lowest beds of the Lias (Pre-planorbis Beds), which do not contain them, are of Triassic age. The appearance of these ammonites is not accompanied by an appreciable change in lithology, yet it marks the establishment of a connection with a southern ocean known as the Tethys from which the British area was separated during parts of Triassic time. From that period onwards, successive waves of ammonites entered the Lias seas in large numbers, and spread rapidly over wide areas. Some families survived longer than others, but all include a variety of species, many combining a short vertical range with wide distribution, features which make them ideal 'zonal' fossils.

The ammonite zones and subzones, of which no fewer than 54 are recognised in the Lias, have proved to be of great value, both in the demonstration of non-sequences and in showing how conditions of deposition varied from place to place at any given period.

A three-fold subdivision of the Lias into Lower, Middle and Upper parts has been made since 1829, although the definition of the term Middle Lias has varied with different workers. During the last hundred years the Middle Lias has most commonly been defined as comprising the *Amaltheus margaritatus* and the *Pleuroceras spinatum* zones, a chronostratigraphical rather than a lithostratigraphical division. However, the Lower/Middle Lias boundary thus defined rarely coincides with a change in lithology and the requisite fossil evidence is not always available. In practice, in the field, the boundary is taken below the Marlstone Rock Bed at a lithological change from sand and siltstone above to mudstone below, which may come within the *margaritatus, davoei* or even the upper part of the *ibex* Zone.

The top of the Lower Jurassic is taken at the base of the *Leioceras opalinum Zone.* In times past this has included the *Pleydellia aalensis* Subzone, but this has now been reclassified as the uppermost subzone of the preceding (Lower Jurassic) *Dumortieria levesquei* Zone. In the southern part of the present region, the topmost two metres or so of the Upper Lias Yeovil and Bridport Sands include part of the *opalinum* Zone and are therefore Middle Jurassic in age, though for convenience of description they are here included with the Lower Jurassic.

LOWER LIAS (including Pre-planorbis Beds)

General description

Although other fossils are common in the Lower Lias, the chronostratigraphy is based almost entirely on the ammonite faunas. The only formal lithostratigraphical units widely recognised within the Lower Lias of the district are the Blue Lias and the overlying Lower Lias clay. A number of informal local units have been recognised in the Radstock area (Table 5) and subdivision of the Blue Lias has been made in the Keynsham – Bath area and along the west Somerset coast where exposure has been, or is very good.

The range of stratigraphical variation in the Lower Lias successions of the district is given in Table 5 and the thickness variation in Figure 22. Fewer details are available in cols. 1 and 5 of the table than for the other columns due to the relative lack of exposure; these sequences may include as yet unidentified, though probably small, nonsequences. The most striking differences are between the Central Somerset (col. 2) and Cotswold (col. 5) deep basinal areas and their flanking positive areas (cols. 3, 6) which show much attenuation and many nonsequences. In the eastern Mendips, between Frome and Binegar, Lias rocks are entirely absent (Figure 22), and the Upper Inferior Oolite rests directly on the Carboniferous Limestone. The Central Somerset Basin continues westwards under the Bristol Channel, where seismic reflection and other evidence indicates very thick sequences, with about 530 m attributed to the combined Lower and Middle Lias and about 90 m to the Upper Lias.

Blue Lias

William Smith adopted the old West Country quarryman's term 'lias' for thin compact limestone beds and distinguished, in a stratigraphical sense, the White Lias below from the darker-coloured Blue Lias above in the Somerset Coalfield. Within the present region, the proportion of limestone to interbedded shale and mudstone varies from a preponderance of limestone in the more condensed sequences, such as in the original type coalfield area, to a ratio of 1:4 or 5 in the thick basinal successions. Wildly varying figures for the thickness of the Blue Lias given by different workers in the past are due mainly to differences of opinion as to the proportion of limestone present needed to define the formation, but also to poor exposure of all except the lowest part in most areas. When long, continuous sections are available, the contrast between the strata containing numerous limestone beds and those containing relatively few beds is usually clear. The Blue Lias commonly extends from the base of the Lias to about midway up the *semicostatum* Zone. In the west Somerset coast sections (Plate 9) it can be shown that the variations of

Stage	Biozone	South Somerset[1]	West Somerset –Glastonbury[2]
Toarcian	*Dumortieria levesquei*	Yeovil Sands (60 – 70 m)	Yeovil Sands (50 – 60 m)
	Grammoceras thouarsense	Junction Bed (1.2 – 4.6 m)	mudstone & silt overlying limestone & mudstone in lower part (50 m)
	Haugia variabilis		
	Hildoceras bifrons		
	Harpoceras falciferum		
	Dactylioceras tenuicostatum		(not proved)
Pliensbachian	*Pleuroceras spinatum*	Marlstone Rock Bed (0.5 – 4 m)	Marlstone Rock Bed (0.5 m)
	Amaltheus margaritatus	sand & silty mudstone (50 – ?90 m)	silty mudstone (70 m)
	Prodactylioceras davoei		mudstone & silty mudstone (40 m)
	Tragophylloceras ibex	'Belemnitiferous Marls' (20 m +)	calcareous mudstone (56 m)
	Uptonia jamesoni		
Sinemurian	*Echioceras raricostatum*	mudstone & shale	
	Oxynoticeras oxynotum	?	mudstone & shale (134 m)
	Asteroceras obtusum	mudstone & shale (?120 m including overlying beds)	
	Caenisites turneri		
	Arnioceras semicostatum	?	Div.5 (part) (16 m)
Hettangian	*Arietites bucklandi*	mudstone with limestone bands (?40 – 50 m)	Div.4 (40 m)
	Schlotheimia angulata		Div.3 (50 m)
	Alsatites liasicus		Div.2 mainly shale (20 m)
	Psiloceras planorbis	mainly limestone (5 – 8 m)	Div.1 (to base of Lias) (8 m)
[Rhaetian]	—	Pre-planorbis Beds (3 – 4 m)	Pre-planorbis Beds (5 m)

(South Somerset and West Somerset columns labelled vertically BLUE LIAS)

† Spiriferina Bed at base   [1] The top of the Lower lias was mapped at the top of the 'Belemnitiferous Marls' on the Yeovil (312) Sheet.   [2] Based mainly on the Brent Knoll boreholes

Table 5   Chrono- and lithostratigraphical classification of the Lower Jurassic and highest Triassic

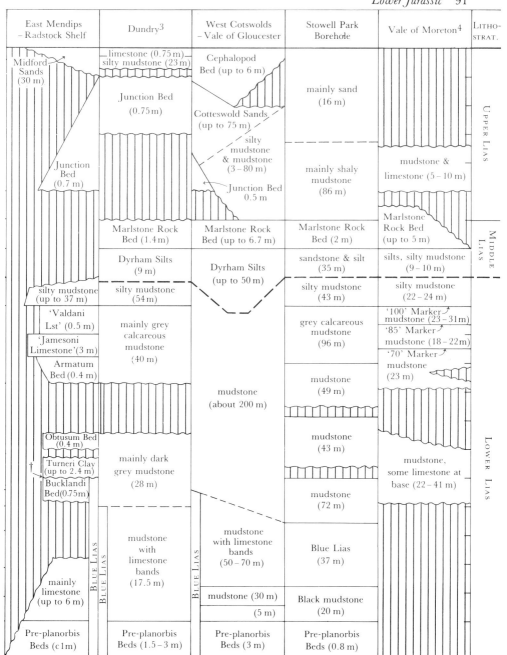

East Mendips – Radstock Shelf	Dundry[3]	West Cotswolds – Vale of Gloucester	Stowell Park Borehole	Vale of Moreton[4]	LITHO-STRAT.
Midford Sands (30 m)	limestone (0.75 m) silty mudstone (23 m)	Cephalopod Bed (up to 6 m)	mainly sand (16 m)		UPPER LIAS
	Junction Bed (0.75 m)	Cotteswold Sands (up to 75 m)			
Junction Bed (0.7 m)		silty mudstone & mudstone (3 – 80 m)	mainly shaly mudstone (86 m)	mudstone & limestone (5 – 10 m)	
		Junction Bed 0.5 m			
	Marlstone Rock Bed (1.4 m)	Marlstone Rock Bed (up to 6.7 m)	Marlstone Rock Bed (2 m)	Marlstone Rock Bed (up to 5 m)	MIDDLE LIAS
	Dyrham Silts (9 m)	Dyrham Silts (up to 50 m)	sandstone & silt (35 m)	silts, silty mudstone (9 – 10 m)	
silty mudstone (up to 37 m)	silty mudstone (54 m)		silty mudstone (43 m)	silty mudstone (22 – 24 m)	
'Valdani Lst' (0.5 m)	mainly grey calcareous mudstone (40 m)		grey calcareous mudstone (96 m)	'100' Marker mudstone (23 – 31 m)	
'Jamesoni Limestone' (3 m)				'85' Marker mudstone (18 – 22 m)	
Armatum Bed (0.4 m)				'70' Marker mudstone (23 m)	
		mudstone (about 200 m)	mudstone (49 m)		LOWER LIAS
Obtusum Bed (0.4 m)			mudstone (43 m)	mudstone, some limestone at base (22 – 41 m)	
Turneri Clay (up to 2.4 m)	mainly dark grey mudstone (28 m)				
Bucklandi Bed (0.75 m)			mudstone (72 m)		
	mudstone with limestone bands (17.5 m)	mudstone with limestone bands (50 – 70 m)	Blue Lias (37 m)		
mainly limestone (up to 6 m)		mudstone (30 m)	Black mudstone (20 m)		
		(5 m)			
Pre-planorbis Beds (c 1 m)	Pre-planorbis Beds (1.5 – 3 m)	Pre-planorbis Beds (3 m)	Pre-planorbis Beds (0.8 m)		

[3] Based mainly on the Dundry Borehole

[4] Based mainly on the Upton and Apley Barn boreholes in the Burford – Witney area

Table 5 *continued*

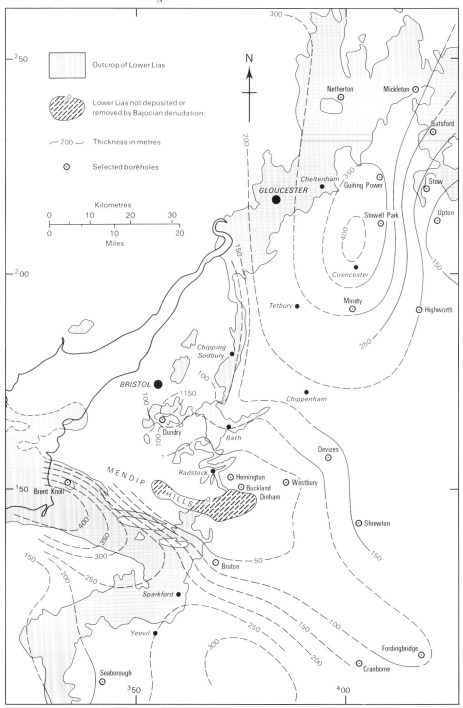

Figure 22   Thickness variations in the Lower Lias. The area between Radstock and Bath (after Donovan and Kellaway, 1984).

Plate 9   Cliffs in Blue Lias (Lower Lias) at Kilve, north Somerset (A11689).

thickness in the Blue Lias are determined by differences in thickness in the mudstones; the intervening limestones retain their identity and thickness over great distances.

Three main types of limestone can be recognised in the Blue Lias. The most common is blue-grey (due to around 25 per cent sulphide-bearing clay content), finely crystalline, hard and splintery, with bedding tops and bottoms either level or wavy. The thicker beds may attain 0.3 to 0.4 m, but the majority are thinner. There is every gradation between the wavy-bedded types, through nodular limestones to layers of limestone nodules. Secondly, beds of flaggy limestone composed of finely comminuted remains of oysters and other bivalves in a limy-mud matrix are commonly present in the Pre-planorbis (or Ostrea) Beds, but in the Mendip–Radstock area this type may also be present at higher horizons. Thirdly, a distinctive type of limestone is commonly represented in the upper part (*johnstoni* Subzone) of the *planorbis* Zone; this has level upper and lower surfaces and is very fine grained, with a porcellanous texture, and may be laminated. The intervening mudstones range from grey, calcareous, blocky, often bioturbated types, to darker-coloured, less calcareous, finely laminated, bituminous types. The latter weather into paper shales and tend to occupy the middle part of the mudstone intervals, and are best represented in the basinal areas.

The cause of the rhythmic alternation of limestone and mudstone has long been debated. Recent opinion favours a cyclical deposition of lime-rich and lime-poor layers, with subsequent diagenetic concentration of the lime in the former to give rise to the majority of the limestones. Hiatuses in sedimentation, which may represent no more than a fraction of an ammonite zone, may be marked by limestone hardgrounds and remanié accumulations of bored and encrusted limestone fragments. These show that limestone formation must have proceeded penecontemporaneously and faunal evidence indicates that a soft muddy bottom was the norm during Blue Lias times so that limestone formation must have proceeded below, rather than on, the sea bed.

Pelagic animals, apart from the ammonites, are represented by belemnites, fishes such as *Pholidophorus* and *Dapedius*, and giant marine saurian reptiles including *Ichthyosaurus* and *Plesiosaurus*. Next to the ammonites, however, the most conspicuous fossils are bivalves, with about 25 genera being represented, including byssally attached, free-swimming and burrowing forms. Brachiopods are locally abundant, especially *Calcirhynchia calcaria* and, in the younger beds, *Spiriferina* and *Piarorhynchia*. Echinoid spines are locally abundant. The prolific faunas appear to reflect rather well-aerated conditions, but fossils may be much reduced in number or absent in the finely laminated beds, which reflect stagnant deep-water conditions. The recovery from time to time of almost complete skeletons of saurians shows that the sea floor must have been below the wave base.

Lower Lias Clay

Exposures are scanty in the Lower Lias Clay and detailed knowledge of these beds depends largely on cored boreholes. In general terms, the lower part consists of dark mudstones and shales with rare, thin, dark argillaceous limestone beds or nodules, which roughly correspond to the Shales-with-Beef and the Black Ven Marls of the Dorset Coast. Next follows a lighter-coloured, more calcareous group, roughly equivalent to the Belemnite Marls of Dorset, ranging in age from within the *raricostatum* Zone through to the *jamesoni* and *ibex* zones. It includes more limestone beds than the underlying darker mudstone group. At the top, rather pyritous, silty and micaceous mudstone, with scattered clay ironstone nodules, characterises the *davoei* Zone and corresponds to the Green Ammonite Beds of Dorset. In most areas the upper beds pass up without appreciable break into the Middle Lias. The fauna of the Lower Lias Clay is again dominated by bivalves, accompanied by brachiopods, ammonites, belemnites and gastropods of considerably greater diversity than in the Blue Lias.

Regional variations

South Somerset

The main part of the Lower Lias outcrop comprises the low-lying clay lands of the vales of Ilchester and Sparkford. These are bounded to the north by the strong north-west-facing escarpment formed by the basal Blue Lias and White Lias limestones, and to the south by the higher ground formed by the Upper Lias sands and the Inferior Oolite. The east–west ridge of Camel Hill, Sparkford, is due to a faulted upfold of Penarth Group to Blue Lias strata. There has been a tendency in this area to limit the term Blue Lias to the flaggy limestone of the Pre-planorbis Beds and the *planorbis* Zone, which were much used in the past for building, paving and monumental stone. Evidence from the adjacent Sparkford railway cutting and elsewhere shows, however, that the strata with numerous limestone beds extend much higher. The lower flaggy limestones can still be seen in quarries on Camel Hill. The remainder of the Lower Lias succession is rarely exposed, but a study of scattered temporary exposures over the years has shown that the ammonite succession is probably more or less complete. The *obtusum* Zone at Marston Magna is notable for its richly fossiliferous nodules packed with *Promicroceras marstonense* and other ammonites. The limestone was formerly cut and polished under the name of

'Marston Marble'. The *ibex* Zone, no more than 2 to 3 m or so thick, at the top of the relatively pale coloured 'Belemnitiferous Marls', probably represents a condensed deposit, as on the Dorset coast.

Central Somerset Basin

This area is flanked on the south by the White Lias – basal Blue Lias limestone scarp known as the Polden Hills and on the north by the Mendip Hills. The centre of the basin contains the thickest known Lower Lias sequence in the region. Much of the area is covered by alluvial deposits of the Somerset Levels, but the beds up to the middle of the *semicostatum* Zone are magnificently exposed in the west Somerset coast sections between Blue Anchor and Hinkley Point. The remainder of the sequence is known in detail from the Burton Row Borehole at Brent Knoll.

The Lower Lias succession can be matched, almost bed-by-bed, throughout the coastal outcrops and with the Brent Knoll succession, a total distance of more than 30 km. The thickness of the strata remains remarkably constant, thinning only in the Blue Ben area, close to the Quantocks, which is believed to have formed a landmass in Lias times. There are, however, no littoral deposits preserved here.

Five divisions in the Lower Lias have been mapped in the coastal area (Table 5, col. 2) using distinctive limestone marker beds. The limestones in the Blue Lias are most abundant in Divisions 1 and 3. Division 2 is mainly shale and mudstone, and approximately corresponds to similar strata recognised in the Bristol area and South Wales. At the base of the Lias, paper shales, nearly 2 m thick in the Watchet area, were formerly classified as 'Watchet Beds', but are now included on lithological grounds with the overlying beds. Inland exposures in the Lower Lias are mainly flaggy limestones in the lower part of the Blue Lias (Division 1 of the coast). These were much quarried in former times along the Polden Hills, between Dunball and Keinton Mandeville. Westwards from Brent Knoll, even greater thicknesses of Lower Lias are present beneath the floor of the Bristol Channel than were proved in the Burton Row Borehole (see p.89).

Mendips – Radstock – Broadfield Down

In early Lias times much of the Mendips formed a large island in a string of smaller islands, including Broadfield Down, stretching westwards into South Wales. The land was fringed in places by a sublittoral zone, 1 to 3 km wide, where the Lias directly overlies the Carboniferous Limestone (Plate 10). The sublittoral facies, known informally as Downside Stone in the Mendips and Brockley Down Limestone on Broadfield Down, consists of pale-coloured, massive, coarse, shelly, and commonly pebbly limestones in which the pebbles and much of the coarse debris are derived from the Carboniferous Limestone. The rocks were deposited in offshore shoals in strongly agitated water. The thickest development, around 30 m, is in the Shepton Mallet area. The rocks are mainly Hettangian in age, but as the old coastline is approached near Maesbury Castle, lateral passage of the facies extends up into the *jamesoni – ibex* zones. The southwards passage of the sublittoral facies into the Blue Lias of the Central Somerset Basin is fairly rapid, whereas north of the Mendip island, the rocks pass into the Radstock facies, representing sedimentation in shallow water on the Radstock Shelf. The limits of the shelf correspond approximately to the 50 m isopach in Figure 22. Tutcher and Trueman recognised eight subdivisions below the clays of the *davoei* Zone (Table 5, col.3) in this classic area. They are mostly too thin to map, but are useful for correlation

Plate 10   Lower Lias (littoral facies) unconformable on Carboniferous Limestone at Lulsgate Quarry, Lulsgate Bottom, Avon (A10720).

purposes. Apart from beds of Hettangian age, the succession consists of mainly condensed sequences and nonsequences, locally with a wealth of ammonites and other fossils which may be rolled and phosphatised.

Tutcher and Trueman thought that folding had occurred during late Hettangian times along east–west axes and that erosion prior to the deposition of the Bucklandi Bed planed off the crests of the folds. The data, however, are inconclusive and a reconstruction over a wider area than they considered gives a different interpretation (Figure 23). Although erosion undoubtedly occurred, it is now thought that most of the differences in thickness below the erosion surface are depositional rather than erosional. The Armatum Bed is the lowest part of the Jamesoni Limestone and, like it, consists of cream-coloured, fine-grained limestone speckled with 'ironshot' (ferruginised fossil fragments). The Valdani Limestone is a coarsely crystalline rock, ironshot at the base. These limestones are limited to the central parts of the Radstock Shelf (Figure 23); elsewhere they pass laterally into grey calcareous mudstones. The silty mudstones of the succeeding *davoei* Zone, which mark the end of the peculiar conditions of the shelf, occur in the Lias successions throughout the region.

The course of sedimentation over most of the Mendips and Broadfield Down in later Liassic times is little known because of the subsequent removal of the deposits. The presence of scattered Lias fissure deposits in the Carboniferous Limestone, however, shows that the sea must have been steadily encroaching upon the land areas, and it is probable that by the end of the period the greater part of the area was submerged.

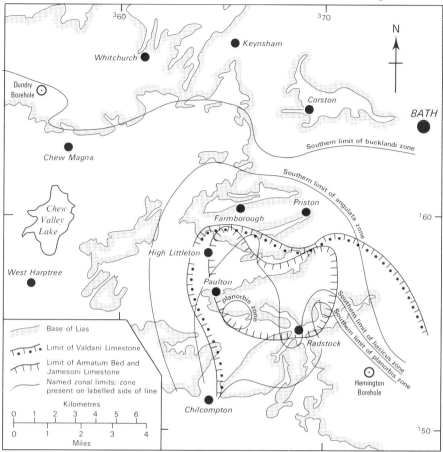

Figure 23 Sketch map illustrating the limits of the *planorbis, liasicus, angulata* and *bucklandi* zones, and the Armatum Bed/Jamesoni Limestone and Valdani Limestone in the area of the Radstock Shelf (after Donovan and Kellaway, 1984, figs 6 and 7).

Bristol – Severn area

The succession given in Table 5 (col. 4) refers mainly to Dundry Hill, where it is most complete and probably thickest (Figure 22). The Blue Lias thickness is comparable to that of the type area at Saltford and Keynsham, although the proportion of interbedded argillaceous strata at Dundry is rather higher. In the type area the Blue Lias has been subdivided into four divisions (A – D). The Saltford Shales (B) comprise part of the *planorbis* and most of the *liasicus* zones. The uppermost division (D) is marked at the base by the Calcaria Bed, a prominent limestone with abundant *Calcirhynchia calcaria* and vermiceratid ammonites, and at the top by the Scipionianum Bed, a thick limestone with phosphatised fossils including *Arnioceras* spp. on its upper surface. These subdivisions can be recognised over quite a wide area, although only the Saltford Shales are mappable. This latter unit thickens from 5 m in the Keynsham area to about 12 m at Chipping Sodbury.

Outliers of Lower Lias occur on either side of the River Severn upstream as far as Chepstow, but only those immediately north of Bristol include strata as late as the *semicostatum* Zone. The subdivisions of Blue Lias at Keynsham can be broadly recognised but, with the exception of Division A (Pre-planorbis Beds, *planorbis* Zone), the strata become markedly more argillaceous.

Cotswolds and Vale of Gloucester

The Stowell Park Borehole between Northleach and Cirencester has provided detailed information of the Lower Lias at its maximum development in the Cotswold Basin (Table 5, col. 6). Compared with the Central Somerset Basin, the Blue Lias includes fewer biozones and the thickness of the individual limestones is less. In contrast to other parts of the region, the Pre-planorbis Beds and the *planorbis* Zone are entirely in a very dark almost black, shaly mudstone facies which forms paper shales at outcrop, and which extends northwards beyond the confines of the district.

When traced westwards away from the basin centre, the Lower Lias shows a marked decrease in thickness (Figure 22). Throughout the greater part of the Vale of Gloucester, the basal flaggy and shelly limestones of the Pre-planorbis Beds and *planorbis* Zone maintain their distinctive characters, but the overlying part of the Blue Lias, comprising clays with numerous limestone bands and nodule horizons, cannot locally be separately mapped from the Lower Lias Clay. The details in Table 5 (col. 5) mainly refer to recent work in the Tewkesbury area (Worssam et al., 1989). Detailed information about the Lower Lias Clay at outcrop is rarely available, though the presence of many of the ammonite zones and subzones has been confirmed in temporary exposures.

The nature of the eastern edge of the Cotswold Basin is relatively well known, chiefly from borehole information. It is defined by the Moreton 'Axis', across which the Lower Lias shows rapid eastwards attenuation (Figure 22). Study of the zonal distribution shows that much of the attenuation is due to the onlap of progressively younger beds of the Lias onto the pre-Lias strata of the London Platform (Donovan et al., 1979) this onlap is accompanied by persistent, though modest thinning of individual units in the same direction. The transgression was not continuous and regressional stages are marked by nonsequences in the succession of the shelf area (Table 5, col. 7). Thin marker horizons, usually highly calcareous beds, can be recognised in boreholes over many tens of kilometres by their characteristic geophysical log signatures. Small faunal and lithological variations can be traced within each of the markers and these, together with the constancy of the intervening strata, are witness to a remarkable uniformity of sedimentation over the shelf.

MIDDLE LIAS

The Middle Lias comprises two formations, a lower, thick silty and arenaceous group (the Dyrham Silts) and an upper, much thinner, ferruginous shallow-water limestone (the Marlstone Rock Bed). These represent the final parts of a regressive phase in the major sedimentary rhythm that commenced with the thick argillaceous Lower Lias succession.

Apart from the outliers of Dundry Hill and Bitton Hill, where thin Marlstone Rock Bed, with (at Dundry) or without the underlying silts, is present, both formations of the Middle Lias are apparently absent between Bitton and the south side of the Mendips. The silts are very thin between Upper Cheyney and Dyrham, but thicken northwards; the Marlstone Rock Bed is absent or unmappable some considerable distance northwards, to near Hawkesbury. The absence or attenuation of the Middle Lias in these areas was mainly due to intra-Bajocian erosion near the Mendips and intra-Liassic erosion farther north.

Dyrham Silts (formerly 'Middle Lias Silts')

The zonal age of these generally poorly fossiliferous beds in most places probably encompasses the *margaritatus* Zone and varying proportions of the underlying *davoei* Zone. The thickness variations broadly follow those of the underlying Lower Lias, though the changes are less well marked (Table 5).

Around Chard and Ilminster, in the extreme south of the region, the lowest deposits consist of blue-grey, micaceous silty mudstone and silt up to 30 m in thickness. These are succeeded by micaceous silt and fine yellow sand, with occasional doggers, known as the Pennard Sands. In the Yeovil district and at Pennard Hill in central Somerset, there is a gradual passage from silty beds up into the Pennard Sands, here 10 to 25 m thick. The thickest recorded sequence of Dyrham Silts (86.8 m) is in the BGS borehole near Bruton.

At Dyrham, the Dyrham Silts are about 20 m thick, and are thought to belong mainly to the *davoei* Zone. Northwards from here, they steadily increase to about 75 m in the mid-Cotswolds. A complete section has been recorded from the now disused Tuffley Brickpit, Robins Wood Hill, south of Gloucester. Here the Marlstone Rock Bed sharply overlies about 58 m of mainly grey silty shales with scattered ironstone nodules. The upper 24 m are assigned to the *margaritatus* Zone and the remainder to the *davoei* Zone. The Lower Lias Clay, belonging to the *ibex* Zone, was represented in the pit bottom by grey clay with calcareous nodules. An almost identical succession was described at the former Stonehouse Brickpit near Stroud. In both pits three beds (0.3 to 1.3 m) of fossiliferous, ferruginous limestone, similar to the later Marlstone Rock Bed, occur within the top 16.5 m of the silts. The uppermost of these beds was also recognised in the Stowell Park Borehole, where it consisted of dark green, calcareous and sideritic, 'chamosite' oolite, about 2 m thick. Similar ferruginous and calcareous beds have been described from areas to the north of the region.

Marlstone Rock Bed

The Marlstone Rock Bed marks a widespread change in conditions of sedimentation. It is typically a shelly, ferruginous, locally oolitic limestone, which may pass into a calcareous ferruginous sandstone. Fossils are abundant and include several species of the ammonite *Pleuroceras*, and numerous belemnites and bivalves; brachiopods such as *Tetrarhynchia tetrahedra* and *Lobothyris punctata* commonly occur in large numbers.

The Marlstone Rock Bed is rarely more than 6 m thick and is only 0.3 m thick at Yeovil on the Yeovil 'high', over which stratal attenuation is known to have occurred at intervals up to and including late Bajocian and early Bathonian times (p.123). In south Somerset, the thin Marlstone Rock Bed is not mapped separately from the immediately overlying limestones known as the 'Junction Bed'.

In the Ilminster district the top 0.2 m of the Marlstone Rock Bed yields ammonites that indicate the presence of the *tenuicostatum* Zone. Over the remaining part

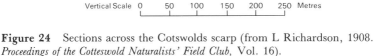

N   S

Leckhampton
Hill

RAGSTONE BEDS
UPPER FREESTONE
OOLITIC MARL
LOWER FREESTONE
PEA GRIT-SCISSUM BEDS
COTTESWOLD SANDS

Hill House

Site of
Pilley
Clay-pit

UPPER LIAS CLAY

Leckhampton
Station
Clay-pit

MARLSTONE
MIDDLE LIAS
(SANDS & CLAY)

Superficial Deposits
(Sand, Gravel, etc.)

LOWER LIAS
(CLAY)

SEA
LEVEL

Section of Leckhampton Hill, Cheltenham

N.15°W   S.15°E

Stinchcombe
Hill

INFERIOR OOLITE
CEPHALOPOD BED

Fieldlane
Farm

Yewtree Inn

COTTESWOLD SANDS

UPPER LIAS CLAY
MARLSTONE
MIDDLE LIAS
(SANDS & CLAY)

LOWER LIAS
(CLAY)

SEA
LEVEL

Section of Stinchcombe Hill

Horizontal Scale   0   50   100   150   200   250   300   Metres

Vertical Scale   0   50   100   150   200   250   Metres

Figure 24   Sections across the Cotswolds scarp (from L Richardson, 1908.
Proceedings of the Cotteswold Naturalists' Field Club, Vol. 16).

of Somerset, this zone has not been recognised and a nonsequence is therefore
postulated at the top of the Marlstone Rock Bed.

Unlike the Marlstone of Oxfordshire, the Marlstone Rock Bed of Gloucestershire and Somerset is normally too thin and its iron content too low to warrant its
use as an ironstone. Its pleasant, rusty-brown colour makes it a most attractive
building stone, but it is rather soft and there is much waste in quarrying.

Owing to their relative hardness in comparison with the soft sands and clays
above and below, the Marlstone Rock Bed and the Junction Bed give rise to
characteristic platform topography, both in Somerset and in the Cotswolds. The
Junction Bed platform is well-marked around Ilminster and Corton Denham in
Somerset; the Marlstone Rock Bed platform is well seen in the Glastonbury outlier. In Gloucestershire, the Marlstone Rock Bed forms a conspicuous ledge below
the Cotswold scarp, a feature which is well exhibited near Wotton-under-Edge and

Stinchcombe (Figure 24). It also gives rise to the flat-topped hills of Diston and Dumbleton, and the ledges on Oxenton and Alderton Hills, between Cheltenham and Broadway.

UPPER LIAS

Sedimentation during the lowest two zones of the Upper Lias was on a modest scale throughout the region, but thereafter the pattern was markedly different in the north and south. In the mid-north Cotswolds the next three to four zones are represented by thick arenaceous and argillaceous sequences, whereas the corresponding strata to the south are thin and much condensed. From *dispansum* Subzone times the situation is almost exactly reversed, with a condensed sequence in the mid-north Cotswolds and a thick arenaceous and argillaceous sequence to the south. The condensed sequences are typically represented by fine-grained, pale-coloured and often ironshot 'cephalopod limestones', in which ammonites may be locally very abundant. These are represented by the Cephalopod Bed in the north and the Junction Bed in the south. Superimposed on this sedimentation pattern, the Upper Lias has suffered a more local attenuation in the areas of the Moreton-in-Marsh and Mendip 'axes' due to a combination of sedimentary thinning and strong erosion (Figure 25). The thickest Upper Lias within the region is present in the Cotswold Basin, whose axis stretches between Bredon Hill and Stowell Park with thicknesses of 110 m and 102 m respectively.

Junction Bed

The Junction Bed consists of thinly bedded grey limestone with ferruginous, oolitic and conglomeratic beds, yielding *Harpoceras, Hildoceras, Dactylioceras* and other mid-Toarcian ammonites. The Junction Bed is known in great detail from the Ilminster area in Somerset, where, within a thickness of 5 to 6 m, the abundance of ammonites, including those of remanié origin, has enabled recognition of all the subzones from the upper part of the *tenuicostatum* Zone to the lowest part of the late Toarcian *levesequei* Zone inclusive. The lowest 1.5 m include the well-known paper shales of the 'Saurian and Fish Bed', with nodules containing fish and reptilian remains, and the 'Leptaena Bed', with a remarkable fauna of very small (micromorphic) forms of brachiopods. The *tenuicostatum* Zone, which here is represented by only of 0.2 m of strata, is unproven and probably absent elsewhere in the present region, apart from the Cotswold Basin. Although it is rarely more than 3 to 5 m thick, the Junction Bed is widely distributed, being found in the main escarpment from Doulting to Chard and also in the outliers of Pennard, Glastonbury and Dundry.

In the eastern Mendips, the Upper Lias is missing, partly as a result of the Bajocian Denudation, but reappears south of Bath in the form of impersistent thin marly and ferruginous ironshot limestones resting on the clays of the Lower Lias and overlain by the Midford Sands (Figure 26). The thickness does not exceed 1.5 m and there is much evidence of erosion and reworking. The beds continue as far north as Old Sodbury where 1.5 m of cream-coloured marl, limestone and conglomerate of *falciferum* Zone age are overlain by a sandy pyritous bed with *Hildoceras bifrons*, at the base of the Cotteswold Sands. From this point northward the Junction Bed passes into the sands and clays at the base of the thick Cotswold Basin succession and ceases to be a recognisable lithological unit.

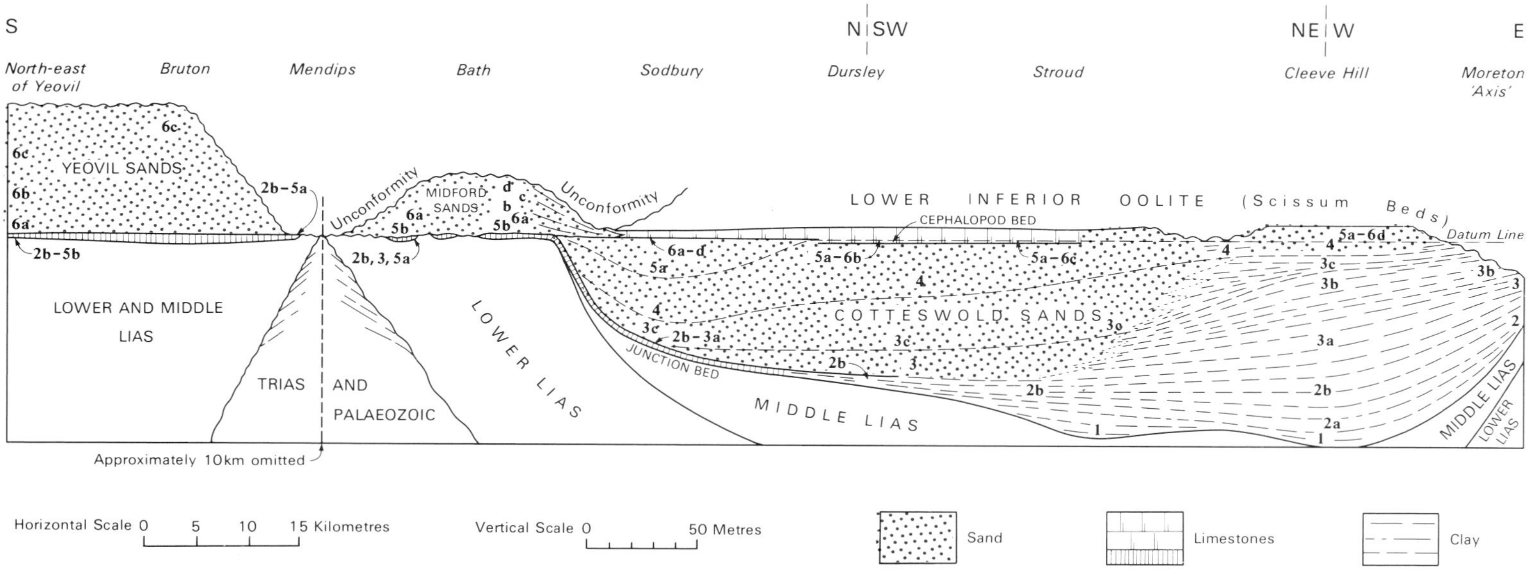

Figure 25   Diagrammatic section of the Upper Lias to illustrate facies changes across the region.

The top of the *striatulum* Subzone (5a) is taken as the datum. Zones and subzones in the figure: *tenuicostatum* (1); *falciferum* (2a, *exaratum*, 2b, *falciferum*); *bifrons* (3a, *commune*, 3b, *fibulatum*, 3c, *crassum*); *variabilis* (4); *thouarsense* (5a, *striatulum*, 5b, *fallaciosum*); *levesquei* (6a, *dispansum*, 6b, *levesquei*, 6c, *moorei*, 6d, *aalensis*).

Upper Lias Sands, Cephalopod Bed and Upper Lias Clay

The lateral passage of the Junction Bed into sand and clay in the south Cotswolds is followed by even more rapid facies changes in the beds above. Before describing these, however, it is necessary to consider the general succession as proved in the escarpment between Old Sodbury and Stroud (see Figures 24 and 25). Here, the sandy limestones of the Scissum Beds (Inferior Oolite) rest upon the Cephalopod Bed. The latter attains its maximum thickness of 4.6 m in the Dursley district, where it consists of ferruginous, oolitic limestones and marls with abundant ammonites and belemnites. The stratigraphical significance of the condensed sequence in the Cephalopod Bed can be interpreted from its fauna which includes many ammonites, e.g. *Grammoceras striatulum, Phlyseogrammoceras dispansum, Dumortieria moorei* and *Pleydellia* (see Figure 25) from which it can be dated as representing the *thouarsense* and *levesquei* zones of the late Toarcian.

Below the Cephalopod Bed lie some 60 m of fine yellow sand with doggers forming the Cotteswold Sands. These in turn rest upon the Junction Bed or on the Upper Lias Clay.

As the Cephalopod Bed is traced southwards from Stroud and Dursley towards Old Sodbury the basal layers, with *Grammoceras thouarsense* and *G. striatulum* (*thouarsense* Zone), become sandy and then thicken as they pass into the Midford Sands (Figure 25). Near Bath the higher parts of the Cephalopod Bed are similarly affected, until in the Midford district, south of Bath, the bulk of the sand falls within the subzone of *Phlyseogrammoceras dispansum* (early *levesquei* Zone). In general, therefore, the sandy facies is older in the north than in the south. To the north of Stroud, the Cephalopod Bed and Cotteswold Sands are seen to die away or pass into clay. This change of facies first affects the lower part of the sands and then spreads upwards into the higher zones so that at Leckhampton Hill, the Cotteswold Sands have practically disappeared (Figure 25). Farther north, in the Cleeve Hill area, the whole succession from the top of the Marlstone Rock Bed to the base of the Scissum Beds is represented by Upper Lias Clay. This formation attains a maximum thickness of 110 m in the Bredon Hill outlier.

The basal beds, comprising parts of the *tenuicostatum* and *falciferum* zones, are thickest in the axial region of the Cotswold Basin and have long been known from the Dumbleton–Gretton area. Here paper shales up to 7 m in thickness contain micromorphic brachiopods ('Leptaena Bed') and laminated limestone nodules ('Saurian and Fish Bed') with well-preserved fish and insect remains. The occurrences are similar to those at Ilminster (see above).

Borehole information downdip shows that the easterly change to a predominantly argillaceous succession takes place along a south-south-east-trending line extending from Leckhampton Hill to a few kilometres east of Malmesbury, and possibly as far as Salisbury Plain. West of Bath, the Midford Sands pass into silty mudstones with *Dumortieria* as shown in the Dundry Hill outlier (Table 5, col.4; Figure 26).

Owing to the absence of the Upper Lias over the Mendips area due to erosion, these facies changes cannot be traced continuously into Somerset. A recent deep borehole near Devizes has, however, provided information at depth east of the Mendips. This shows a succession of 'southern aspect' with 38 m of Midford Sands overlying 30.5 m of clay and finally 6 m of Junction Bed facies. The older borehole at Westbury (Wilts), much quoted in this context (for instance, Kellaway and Welch 1948, p.56), is now thought, on seismic reflection evidence, to pass through the Vale of Pewsey Fault, thereby cutting out much of the Lias succession.

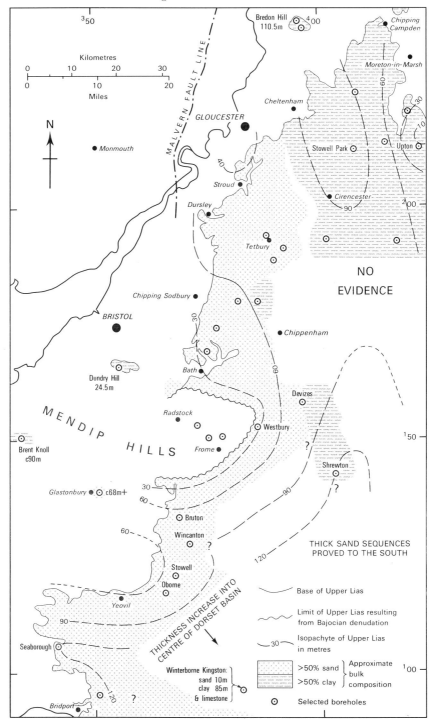

Figure 26   Sketch map showing thickness variations and dominant lithological composition of the Upper Lias.

South of the Mendips, the arenaceous facies, known as the Yeovil Sands (Figure 25), is well developed along the outcrop throughout the district and continues southwards, as the Bridport Sand, to the Dorset Coast. The thickness averages about 60 m and is closely comparable with the Midford Sands. Typically it consists of fine-grained, buff yellow, friable sandstone with layers of calcareous sand-'burrs', which contain comminuted shell debris. Well-preserved fossils are rare in the Yeovil Sands, but evidence of the *dispansum* Subzone of the *levesquei* Zone is found at Barrington near Ilminster. The main mass probably belongs to the *levesquei* and *moorei* subzones of the *levesquei* Zone. Most of the best-preserved fossils come from the upper part of the succession and include *Dumortieria moorei*, *Trigonia charlockensis* and *Rhynchonelloidea cynica*. The youngest sands occur in the south-west part of the district, west of Yeovil Junction, where thin representatives of the *aalensis* Subzone and the succeeding *opalinum* Zone (Middle Jurassic) are usually present. These form part of the Inferior Oolite in the Cotswolds. Northwards from Yeovil, these higher beds appear to have been removed and the Inferior Oolite rests unconformably on the *moorei* Subzone. In the Yeovil area, the lower part of the formation consists of small interdigitating lenses of impure limestone, blue-grey micaceous silty mudstone and pale grey silt. These pass upwards into strata which contain increasing quantities of silt and fine sand, while the top 15 to 20 m, the only part to merit the name of Yeovil Sands, are of the typical sand facies described above. At Ham Hill, near Yeovil, a large mass of shelly limestone of the type present as small lenses or beds elsewhere is known as the 'Ham Hill Stone'. This great lens of shelly debris held together by a ferruginous cement is up to 27 m thick and passes laterally into sand with calcareous lenticles. It was extensively quarried at Ham Hill for use as a building stone.

West of the main outcrop, the two outliers of Glastonbury Tor and Brent Knoll provide evidence for the westwards replacement of sand by clay and silt, already noted at Dundry, north of the Mendips. At Glastonbury, underneath 53 m of sands (top not seen), about 11 m of clays supervene above the Junction Bed. Farther west, at Brent Knoll, the sands have passed into silts, and the underlying beds comprise mudstones with pale grey rubbly and nodular limestone beds in the lower part (Table 5, col.2). Offshore to the west, in the Bristol Channel, the entire Upper Lias sequence is apparently argillaceous.

Downdip from the main outcrop to the south-east, and beyond the confines of the region, deep boreholes drilled in the last decade or so have proved a thickening of the Upper Lias into the Wessex Basin similar to that already noted at Devizes. Both the upper sands and the lower clays share in the thickness increase.

Provenance and distribution of the sands

Following Buckman's (1889) classic demonstration of the southwards younging of the sands in the Upper Lias, which suggests derivation from the north-east, and Boswell's study (1924) of the heavy mineral content, which supports derivation from the Armorican metamorphic terrain of the Brittany area to the south-west, the mode of sedimentation of the sands has provided a puzzle. Palaeocurrent analysis based on cross-bedding measurements (Davies, 1969) showed a dominant current flow to the south-west and a subordinate flow to the north-east.

In the most detailed explanation so far advanced, Davies (1969) postulated a separation of the processes of sediment supply and sediment deposition. Supply from the Armorican source area was by means of longshore drift to the north adjacent to a Cornubian land mass, lying some distance to the west of the present sand outcrop. Deposition was by means of a sediment return system activated by the dominant internal basinal current flow which was towards the south-west, as shown by the cross-bedding measurements. He postulated that the return flow was stimulated by a marked change in the orientation of the 'Cornubian' coastline west of Cheltenham. The sand body itself represents an east–west-trending bar complex that moved southwards from the vicinity of Cheltenham to Bridport by means of accretion of sediment on its southern side. Forebar as well as backbar sands are claimed to be recognised. The Ham Hill Stone which, unlike the main sand mass, shows cross-bedding indicating a strong counterflow direction to the north-east during *moorei* subzonal (late *levesquei*) times is said to represent the deposits of a tidal channel cut through the bar.

This model has been criticised on the grounds that, apart from the Ham Hill Stone, the supposed bar deposits do not show large-scale cross-bedding and that there appears to be no significant difference between the fore- and backbar deposits. In addition, the only surviving deposits to the west are silts and clays (Figure 26). A more recent suggestion, that the sands entered the region from the north-east via the western margin of the London Platform, encounters a similar difficulty in that the central Cotswold clay basin apparently intervenes (Figure 26).

11   Middle Jurassic (Inferior Oolite Group)

The Upper Lias is succeeded by a sequence of shallow-water marine limestones, known as the Inferior Oolite Group, first identified by William Smith. The group extends through the Aalenian and Bajocian stages into the earliest part of the Bathonian stage.

Throughout Gloucestershire the Inferior Oolite forms the great indented scarp of the Cotswolds Hills, which forms a natural boundary between the Severn Valley to the west and the limestone dip slope of the Cotswolds to the east. From Bath to Doulting the Inferior Oolite forms an elevated tract of dissected country that passes athwart the eastern end of the Mendips, and then crowns a low range of hills extending southwards to Sherborne. Thence, the escarpment continues west-south-westwards to Yeovil, beyond which it rapidly diminishes in the belt of attenuated, condensed and much-faulted limestones that runs westwards to Crewkerne and Chard.

East of the main outcrop, the Inferior Oolite underlies the remainder of the district and over the last few decades deep boreholes have much extended our knowledge in these areas. The most westerly onshore outcrop of the Inferior Oolite forms the capping of Brent Knoll where only the Lower Inferior Oolite survives.

CLASSIFICATION

The early classifications of the Inferior Oolite were based on lithology and mixed faunal assemblages. At around the turn of the century S S Buckman, in a series of classic papers, showed that the previous work had failed to recognise the complexity of the stratigraphy and that the key to its understanding lay in the ammonite sequences, supplemented in the Cotswolds, where ammonites are comparitively rare, by the use of brachiopods. This pioneer work, which was greeted with much hostility and misunderstanding, has since formed the basis of our present, but still incomplete understanding and classification of these rocks.

Table 6 shows present-day usage, in which the Aalenian is represented as a separate stage rather than as the lowest part of the Bajocian, formerly the favoured British practice. The ammonite zonal scheme, though extensively revised, must to some extent be regarded still as being provisional. The position of the Middle/Upper Inferior Oolite boundary coincides with the top of the *Strenoceras subfurcatum* (formerly *S. niortensis*) Zone, following the recognition that the lowest beds of the Upper Inferior Oolite, attributable to the *garantiana* Zone, rest unconformably or in nonsequence, as originally postulated by Buckman, on the beds beneath.

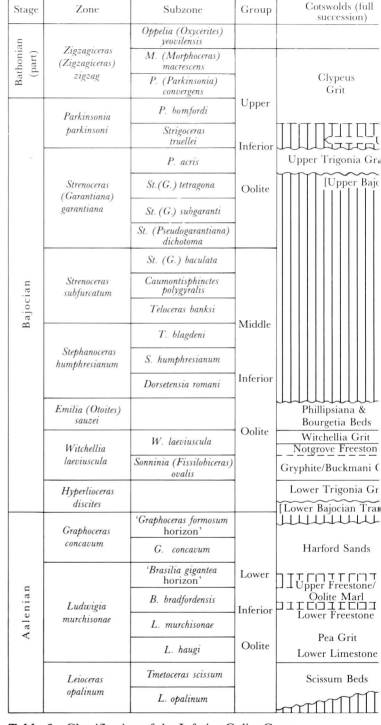

Stage	Zone	Subzone	Group		Cotswolds (full succession)
Bathonian (part)	Zigzagiceras (Zigzagiceras) zigzag	Oppelia (Oxycerites) yeovilensis			Clypeus Grit
		M. (Morphoceras) macrescens	Upper		
		P. (Parkinsonia) convergens			
Bajocian	Parkinsonia parkinsoni	P. bomfordi		Inferior	
		Strigoceras truellei			
	Strenoceras (Garantiana) garantiana	P. acris			Upper Trigonia Gr.
		St.(G.) tetragona	Oolite		[Upper Baj.
		St. (G.) subgaranti			
		St. (Pseudogarantiana) dichotoma			
	Strenoceras subfurcatum	St. (G.) baculata			
		Caumontisphinctes polygyralis			
		Teloceras banksi	Middle		
	Stephanoceras humphresianum	T. blagdeni			
		S. humphresianum		Inferior	
		Dorsetensia romani			
	Emilia (Otoites) sauzei		Oolite		Phillipsiana & Bourgetia Beds
	Witchellia laeviuscula	W. laeviuscula			Witchellia Grit
					Notgrove Freeston
		Sonninia (Fissilobiceras) ovalis			Gryphite/Buckmani (
	Hyperlioceras discites				Lower Trigonia Gr
					[Lower Bajocian Tra
Aalenian	Graphoceras concavum	'Graphoceras formosum horizon'			
		G. concavum			Harford Sands
	Ludwigia murchisonae	'Brasilia gigantea horizon'	Lower		Upper Freestone/ Oolite Marl
		B. bradfordensis		Inferior	
		L. murchisonae			Lower Freestone
		L. haugi	Oolite		Pea Grit
					Lower Limestone
	Leioceras opalinum	Tmetoceras scissum			Scissum Beds
		L. opalinum			

Table 6   Classification of the Inferior Oolite Group

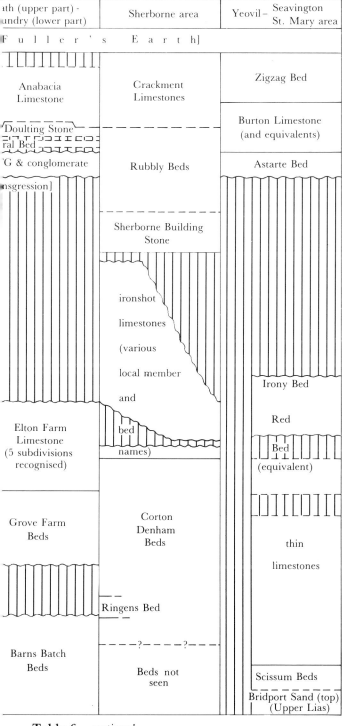

Table 6 *continued*

DEPOSITIONAL PATTERN

The general characters of the Inferior Oolite indicate that most of the succession was formed in a shallow shelf sea, where deposition was interrupted and/or modified at frequent intervals by earth movements, which caused slight warping of the sea floor, or by changes in sea level. Nondeposition, or erosion of the sediments, sometimes before consolidation, resulted from the sea bed coming within the influence of current action. Although these events were not big enough to cause visible angular discordance at any one outcrop, they often led to appreciable gaps or nonsequences in the succession, some of which can be recognised over very wide areas. Where subsidence or a rise in sea level followed, there was commonly overstep of older by younger rocks.

The presence of a nonsequence is often, but not invariably indicated by the occurrence of planed, bored and oyster-covered surfaces, by limestone conglomerates, or by condensed deposits in which fossils of more than one faunal horizon are to be found. Some fossils are often rolled and have coatings of limonite or phosphatic material, or encrusting serpulids.

The isopach and distribution map (Figure 27) of the Inferior Oolite clearly shows two main areas of thick deposits, located in the Cotswolds and in south-east Somerset, and separated by a wide belt of relatively thinner deposits in which the major part or the whole part, of the sequence comprises the Upper Inferior Oolite. The successions in these two areas differ markedly. In the Cotswolds, the thickness of the Upper Inferior Oolite is relatively constant; thus variations of the Inferior Oolite as a whole indicate changes in the residual thicknesses of the Lower and Middle divisions. The depositional environment was dominantly one of very shallow-water, high-energy conditions. The limestones are predominantly oolitic and shell-fragmental, and contain relatively few, poorly preserved ammonites. The Lower, Middle and Upper divisions are separated by well-marked unconformities.

By contrast, in the southern area, the Upper Inferior Oolite shows considerable thickness variations, which follow those of the lower subdivisions. The environment was, on the whole, one of deeper water and lower energy conditions than in the north. The clear distinction between Lower and Middle Inferior Oolite is no longer present. The limestones are mostly finer grained and more marly, and are ferruginous at certain horizons, with abundant geothite ooliths ('ironshots'), pellets and concretions and incrustations of various sorts. Ammonites are abundant and well preserved.

At outcrop, the only reasonably full development of the southern sequence occurs in the Sherborne–Milborne Port area (Table 6, col.3). The thickest sequences occur farther to the east in the subcrop.

This 'basinal' area was flanked on the west and south by a broad positive belt, known as the South Dorset High (Figure 27), in which sedimentation was much reduced and the succession very variable and highly condensed locally; the Lower and Middle Inferior Oolite may be absent altogether (Table 6, col.4). The ferruginisation of the limestones is more widespread and better developed here than in the adjacent basinal area.

The passage of the 'northern facies' into the 'southern facies' occurs in the Upper Inferior Oolite in the Wincanton area and appears to be related to the development of the Mere Fault. It is not known where the corresponding facies change occurred in the Lower and Middle Inferior Oolite because of the extensive erosion that these beds suffered in earliest Upper Bajocian times. It must, however, have

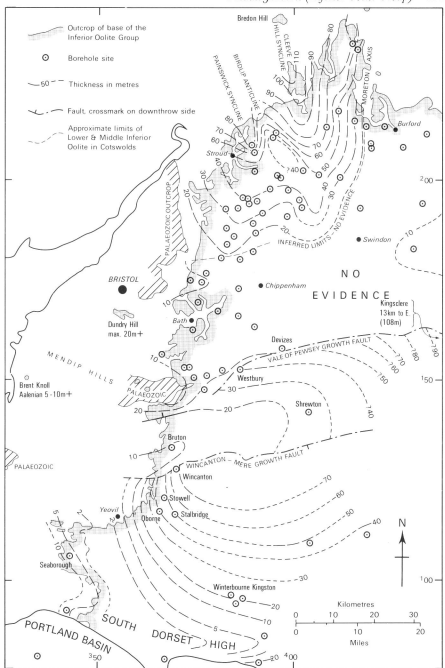

Figure 27 Isopach map of the Inferior Oolite Group.

been a considerable distance to the north of where it lay in Upper Inferior Oolite times, because the deposits on Dundry Hill are of 'southern facies'.

Kellaway and Welch (1948) suggested that the Inferior Oolite suffered attenuation along a narrow north-south belt (the 'Bath Axis'), that represented the southern prolongation of the Malvern 'line' to the Dorset Coast at Burton Bradstock. Subsequent drilling results, however, indicate that south of the Mendips this attenuation appears to be related to broadly east–west trends in the Wessex Basin (Figure 27).

LOWER AND MIDDLE INFERIOR OOLITE

COTSWOLD AREA

S S Buckman, at around the turn of the century, subdivided the Middle Inferior Oolite and defined the relationship of the Lower, Middle and Upper parts to each other. In the Cotswold area (Table 6, col. 1), the Lower Inferior Oolite is primarily an oolite freestone; the Middle Inferior Oolite has been described as 'Ragstone Beds'. The use of the term 'grit' for some of the members is misleading because the rocks, though comprising hard, shelly and rubbly limestones with a fine-grained 'gritty' bioclastic matrix, lack any coarse quartzose sand or 'grit' content.

The Lower Inferior Oolite of the area has recently been re-examined by Mudge (1978), who proposed a new nomenclature, to which, however, some objection has been raised (Cope et al., 1980). The following description takes cognisance of the former account and the new names are added in parenthesis where appropriate. For purposes of description the long established names are retained, but the new work enables them to be applied with a precision and consistency that was often lacking in the older accounts.

Scissum Beds (Leckhampton Limestone)

These yellow to brown, rather fine-grained, sandy limestones and calcareous sandstones extend throughout the area shown as Lower and Middle Inferior Oolite (Figure 27). Although the Scissum Beds usually rest with nonsequence on the Upper Lias and locally include a basal conglomerate, the lower limit may, in the absence of exposures, be difficult to locate where they directly overlie Cotteswold Sands. The rocks are bioturbated and their somewhat sparse fauna is mainly of bivalves and rhynchonellids, especially *Homoeorhynchia cynocephala* and *Rhynchonelloidea subangulata*. The upper part is appreciably more calcareous than the lower and may contain limonitised pellets and scattered ooliths. The thickness is rarely more than a few metres.

Lower Limestone

In the Stroud-Wotton-under-Edge area, the Lower Limestone comprises up to 10 m of unfossiliferous, massive, current-bedded oolite (Frocester Hill Oolite) which rests on a planed, oyster-encrusted surface of the Scissum Beds. It may include an appreciable coarse sand fraction and, locally, even rounded grey quartzite pebbles, as in the Stroud district, where it is termed 'Dapple Beds'. It passes eastwards and northwards into well-bedded, oolitic, bioclastic limestones (Crickley Limestone) that show a downward passage into the underlying Scissum Beds.

Pea Grit

The beds which give this member its name are the very coarsely pisolitic limestones of the Crickley Hill(Plate 11A) – Leckhampton Hill – Cleeve Hill sections (Crickley Oncolite). The distinctive pisoliths are disc-shaped, algal-coated grains (oncoliths) in which the micrite coatings include moulds of algal tubes and filaments. The beds, which measure about 2 to 5 m in thickness, are buff, marly and rubbly with an abundant fauna in which micromorphic brachiopods, gastropods and regular echinoids are notable. Bivalves are abundant and the normal-sized brachiopods include *Stroudithyris pisolithica*, *Plectothyris plicata* and *Epithyris submaxillata*.

The term Pea Grit, or Pea Grit 'Series', has generally been extended to include the variable group of rubbly, pelloidal, coralliferous and sparsely pisolitic limestone and interbedded oolitic rocks (Fiddler's Elbow Limestone) that intervene between the main pisolite (Crickley Oncolite) and the typical oolite-freestones of the succeeding member. Mudge considers that this variable group is replaced laterally by a much bioturbated, dominantly oolitic facies (Cleeve Hill Oolite) in the Cheltenham area, which the older accounts generally include in the Lower Freestone. The Pea Grit, in its wider sense, measures about 8 to 12 m in thickness and is recognised at outcrop between Cleeve Hill and the Stroud area; it continues in a south to south-easterly direction beneath the cover of younger rocks. To the west of this belt, the greater part of the Lower Inferior Oolite comprises cross-bedded oolites which cannot be subdivided with certainty, while eastwards the Pea Grit apparently passes laterally into a warm brown freestone known as 'Guiting Stone' (Jackdaw Quarry Oolite), which has been extensively worked at Temple Guiting and Stanway on the western margin of the Cotswolds, and to the east at Bourton-on-the-Hill. This rock is known informally as 'Yellow Guiting', the overlying 'White Guiting' being the equivalent of the Lower Freestone.

Lower Freestone (Devil's Chimney Oolite)

The Lower Freestone is the thickest and one of the most distinctive members of the Lower Inferior Oolite. It is important in Cheltenham as a building stone and was once extensively quarried on Leckhampton and Cleeve hills, where its thickness and massive, uniform oolitic nature and relative freedom from shells made it ideal for dimension stone and carving. The beds are strongly current-bedded. It is said by Mudge (1978) to be nearly 40 m thick at Leckhampton, where most of the underlying Cleeve Hill Oolite is included. Recent BGS work at Cleeve Hill, however, shows that this section is faulted. A borehole on Cleeve Cloud has proved a total thickness, together with the Cleeve Hill Oolite, of 51 m. The Lower Freestone has long been worked under the name 'Campden Stone' on the hill above Chipping Campden. It is still worked at Westington Hill Quarry and, farther south-west, in Jackdaw Quarry at Stanway (see above).

Oolite Marl – Upper Freestone (Scottsquar Hill Limestone)

Recent detailed work confirms Buckman's original contention that the Oolite Marl and Upper Freestone represents a single sedimentary unit of interdigitating facies (Baker, 1981). This highly distinctive member rests nonsequentially on the Lower Freestone, whose upper surface is an oyster-encrusted, planed and bored hard-ground, although the time gap represented is probably quite short. Three main

Plate 11   Middle Jurassic limestones
A. Inferior Oolite at Crickley Hill, Gloucestershire (A10947)
B. Great Oolite at Avoncliffe, Wiltshire. Twinhoe Beds resting on Combe Down
Oolite (A9734).

facies are represented. First, a marl-dominated 'trough' facies, corresponding to the Oolite Marl, which comprises marls with subordinate calcite-mudstones (micrites) that may contain abundant pellets (pelmicrite). Second, an oolite-dominated 'shoal' facies which comprises ooliths in a matrix ranging from calcite mud (oomicrite) to clear calcite (oosparite), which corresponds to the Upper Freestone. Finally, a mainly micritic marginal facies which shows a range of lithologies transitional to the two main types and dependent for its characters on proximity to either shoal or trough deposits; this facies shows evidence of much reworking of the sediments. The finer-grained rocks are often highly fossiliferous. Brachiopods are particularly abundant and, as originally recognised by Buckman, show stratigraphical zonation. The latest work distinguishes a lower *Zeilleria – Flabellirhynchia lycetti* fauna, associated with micromorphic brachiopods, and an upper *Globirhynchia – Plectothyris fimbria* fauna. Reworking has led to a mixture of forms in a few localities.

Although the oolite facies overlies the marl facies over the greater part of the area, this is not always so, for the oolite dominates the entire succession in the Stroud area, and the marly-micritic facies dominates in the area of the Painswick and Cleeve Hills troughs (see below). The best exposure of the beds is in Westington Hill Quarry above Chipping Campden.

Harford Sands, Snowshill Clay and Tilestone

This variable group of beds occupies a restricted area in the north Cotswolds and is best seen in the region between Winchcombe and Broad Campden. Although the members are described as an ordered tripartite sequence, this is only an approximation and the members should more accurately be regarded as facies types.

The lowest member, the Harford Sands, consists of up to 3 m of pale brown quartz sand, which is frequently hardened to form doggers or sand-burrs. Mineralogically, they are characterised by abundant grains of sphene and rare kyanite. The sands were formerly dug on Cleeve Hill and carried to the Staffordshire potteries.

The Snowshill Clay comprises some 4 to 5 m of stiff, chocolate-coloured clay, which reaches its maximum thickness at Blockley. The overlying Tilestone is also best developed in the Blockley area, where it consists of sandy, oolitic, flaggy limestone, locally containing rolled pebbles of oolite.

Lower Trigonia Grit

This basal member of the Middle Inferior Oolite was deposited upon the eroded surface of underlying strata, and its base is frequently conglomeratic. The deposit consists of rubbly, commonly ironshot limestone crowded with fossils, particularly bivalves such as *Trigonia*. A coral bed with *Latomeandra* occurs near the base.

Buckmani and Gryphite Grits

This block of yellow and brown, sandy, marly rubbly, shelly and commonly ironshot limestones was divided by Buckman into a lower part characterised by *Lobothyris buckmani* and an upper part with an abundance of the oyster *Gryphaea*.

Notgrove Freestone

The Notgrove Freestone has a wide extent in the north and central Cotswolds, where it attains a thickness of 4.5 to 8 m. It consists of hard, white, fine-textured oolitic limestone, locally crowded with shells of *Propeamussium* cf. *laeviradiatum* (formerly *Variamussium pumilum*).

Witchellia Grit

The Witchellia Grit consists of thin, grey-brown, ironshot limestone containing ammonites in greater abundance and in a better state of preservation than the lower beds.

Phillipsiana and Bourguetia Beds

These beds, representing the highest part of the Middle Inferior Oolite preserved in the Cotswolds, are confined to the Cleeve Hill Syncline. They consist of hard, shelly limestones yielding '*Terebratula*' *phillipsiana* and the gastropod *Bourguetia striata.* Many of the brachiopod shells are beekitised (silicified).

The Bajocian transgressions in the Cotswolds

During Bajocian times the Cotswold area was affected by two main periods of earth movement and their associated intraformational erosion and transgression. The Lower, Middle and Upper Inferior Oolite deposits reflect the effects of the movements.

Deposition of the Inferior Oolite in the Cotswolds commenced with the formation of the Scissum Beds and continued without major interruption until the close of Lower Freestone times. Subsidence during this period appears to have been greatest in the Cheltenham–Cleeve Hill area, which formed the centre of a broad basin of deposition where the maximum thickness of the sediments was built up.

After a break in sedimentation at the end of Lower Freestone times, the Upper Freestone and oolite marl were laid down. Facies variation within these rocks provides the first clear evidence of the initiation of the Painswick Syncline or Trough and the Birdlip Anticline or 'High', as well as the continuing development of the Cleeve Hill Syncline. The 'high' areas are distinguished by thinner sedimentation and oolitic shoal facies, the 'low' areas by thicker sedimentation with micritic 'trough' facies (see above). Slight general uplift, followed by a change in conditions, led to the deposition of the Harford Sands, which apparently spread from the east. They were overlapped first by the Snowshill Clay and then by the Tilestones. The full extent of the area of deposition, particularly to the west of the Cleeve Hill Syncline, cannot be estimated owing to the effect of later erosion.

Deposition may have been continuous in the centre of the Cleeve Hill Syncline, though the nature of the sediments suggests that the waters were shallow and landlocked. The deposition of the sandy limestone known as the Tilestone, with its low-diversity marine fauna, took place in more open water. Thereafter, the whole area was submerged beneath the waters of the Lower Trigonia Grit sea. This extension of the area of deposition constitutes the Lower Bajocian Transgression. During deposition of the Middle Inferior Oolite, thickening of the various divisions, as they are traced from Painswick into the Cleeve Hill Syncline, suggests that once again subsidence was greatest in the latter area.

In the Cotswolds, the date of the second, major phase of warping can only be fixed as later than the Phillipsiana Beds and earlier than the Upper Trigonia Grit, since no intermediate strata are represented north of the Mendips. This warping, which was more widespread and of greater intensity than that of the earlier Aalenian – Lower Bajocian movements, completed the shaping of the Painswick and Cleeve Hill synclines and the intervening Birdlip Anticline. Following this episode, the entire Cotswold area was elevated and subjected to erosion and the Middle Inferior Oolite was completely removed from the area of the Birdlip Anticline.

Erosion and planation were followed by subsidence of all the country from the Mendips to Moreton-in-Marsh, and the limestones of the Upper Trigonia Grit were laid down upon the bored and eroded surface of the older rocks.

The form of the denuded folds upon which the Upper Trigonia Grit was deposited in part of the Cotswolds is shown in plan (Figure 28) and cross-section (Figure 36).

DUNDRY AND THE AREA SOUTH OF THE MENDIPS

The main outcrop of the Inferior Oolite from Old Sodbury to Doulting shows Upper Inferior Oolite resting upon an erosional surface of Upper Lias or older strata (Figure 36). At Dundry Hill (Table 6, col.2), however, in the most westerly outlier north of the Mendips, up to about 6 m of Lower and Middle Inferior Oolite are preserved in a shallow syncline beneath the Upper Inferior Oolite in the western half of the hill. The Upper Inferior Oolite rests directly on Upper Lias at the eastern end of the outlier. The close similarity between the Dundry succession and contemporaneous strata in Dorset and east Somerset indicates that these areas were within the same depositional and faunal province during Lower and Middle Inferior Oolite times. Sandy ferruginous beds and hard limestones with limonitic ooliths ('ironshots') are the typical Dundry rock-types. Of the fossils, brachiopods are common, while all groups of the mollusca are abundant, with numerous well-preserved ammonites.

Middle and Lower Inferior Oolite rocks are also preserved south of the Mendips in the shallow Cole Syncline at Bruton, in east Somerset (Figure 36). Here, rocks of the *blagdeni* Subzone are present beneath the Upper Inferior Oolite. These were thought to represent the youngest beds preserved beneath the Upper Bajocian transgression, but in the Doulting area the conglomeratic limestone below the base of the Upper Inferior Oolite (Doulting Stone) is now considered to belong to the younger, lowest subzone of the *subfurcatum* Zone.

There is evidence that the Mendip 'High' and its north-eastward continuation through Trowbridge and thence along the line of important anticlinal structures in the Cretaceous rocks, was bounded to the south by an important growth fault, which was active throughout much of the Mesozoic (see Figure 34). In Bajocian times, the most dramatic known movement along this structure occured at Kingsclere to the east of the present region where one of the thickest known Inferior Oolite successions, some 108 m, occurs a short distance to the south of successions only 10 m or so thick. In the present region this structure is known as the Vale of Pewsey Fault; the Westbury Borehole provides evidence for its activity in Middle Jurassic times. The borehole proved a thick succession of Inferior Oolite (39 m) immediately south of the wide belt in which only relatively thin Upper Inferior Oolite is present (Figure 27). Unfortunately, owing to very poor core

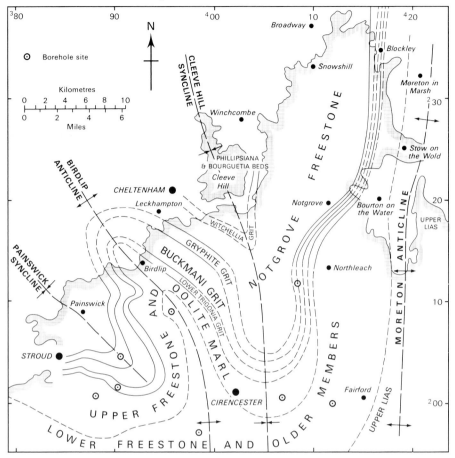

Figure 28   Sketch map of part of the Cotswolds showing the beds upon which the Upper Inferior Oolite was deposited at the time of the late Bajocian transgression.

The stipple encloses the outcrop of the Inferior Oolite and later rocks. The Lower and Middle Inferior Oolite outcrops are shown as they would appear if the Upper Inferior Oolite was removed. (After S S Buckman, 1901 with revisions mainly due to recent boreholes and six-inch primary survey by Cave and Ackerman southwards from Birdlip).

recovery in the borehole, the thickness of Lower and Middle Inferior Oolite is not known, though this must have been considerable.

Southwards from the Cole Syncline, the Mere Fault appears to have acted as a controlling infuence on sedimentation in an analagous way to the Pewsey Fault, with only thin Upper Inferior Oolite to the north of it and a much thicker succession to the south including both the Lower and Middle divisions. At outcrop, the Lower and Middle Inferior Oolite reappear on the south side of the Mere Fault and, within 2 km or less measure 8 to 10 m in thickness. The change is, however, most dramatic at Wincanton (Figure 27) where the borehole adjacent to the fault proved a thickness of at least 40 m of the lower divisions. Within the basin, the uppermost part of the Middle Inferior Oolite comprises a distinctive sequence, no more than 0.5 to 2 m thick, of fossiliferous, ironshot, glauconitic limestones with abundant ammonites. These beds, which contain local nonsequences and condensed successions, span the zonal range of *subfurcatum* to *laeviuscula*. The most complete sequence at outcrop occurs to the east and north-east of Sherborne (Table 6, col.3). These ironshot limestones correspond to the highly condensed 'Irony Bed', often barely more than a few centimetres thick, in the area west of the basin. Within the basin, the ironshot beds overlie fossiliferous, hard grey, variably rubbly and sandy limestone with irregular marly clay partings. These beds are now named the Corton Denham Beds, the upper limit of which is taken at a minor unconformity just above the base of the overlying ironshot beds (Table 6, col.3). The thickness ranges from 6 to 8 m. The Ringens Bed is a widespread fossiliferous marker horizon about 2 m above the base which contains abundant brachiopods, notably *Homoeorhynchia ringens*.

Westwards from Sherborne, the thickness of the Lower and Middle Inferior Oolite diminishes steadily, and the greatly condensed successions thereabouts have long been famous for the profusion of their fossils. S S Buckman's work in unravelling the complexity of the ammonite successions of the area has remained classic. At Halfway House, between Sherborne and Yeovil, these strata comprise 1.7 m of mainly ironshot limestone with the 'Irony Bed' at the top and the Ringens Bed lying only a few centimetres above the base (Table 6, col.4). This much attenuated thickness for the combined Lower and Middle Inferior Oolite persists for nearly 20 km westwards along the outcrop. Locally, in what may informally be called the Yeovil 'High' at Clifton Maybrook and Haselbury Mill, respectively 2 km south-east and 11 km south-west of Yeovil, the beds are absent, and in the Stoford area they range from 0.5 to 0.85 m. Elsewhere within the marginal belt they are only slightly thicker, with the Lower Inferior Oolite being thicker and more widespread than the Middle Inferior Oolite. The latter is absent over most, if not all of the Misterton – Crewkerne – Haselbury area.

Not until the western edge of the Inferior Oolite outcrop near Seavington St Mary and Hinton St George is reached do the strata thicken to any significant degree. Although there are no complete sections, it is probable that the Lower and Middle Inferior Oolite may together attain 3 m or so. Of this thickness approximately the upper third is attributable to the Middle division, which is roughly equivalent to the well-known condensed ferruginous 'Red Bed' of the Dorset coast (Table 6, col.4). Further evidence for this westward thickening is provided by a borehole drilled at Seaborough in 1974 and by outcrops in Dorset, to the south of the present region (Figure 27).

Discussion of the conditions of deposition and sedimentation of the Lower and Middle Inferior Oolite of this southern Dorset – south Somerset area is discussed below (see p.123).

UPPER INFERIOR OOLITE

As a result of the widespread warping and erosion that took place prior to its deposition, the Upper Inferior Oolite rests on formations ranging from the Middle Inferior Oolite to the Lias and, in the eastern Mendips, directly overlies Triassic and Palaeozoic rocks. (see p.117; Figure 36). In the main Dorset – Wessex Basin, the western edge of which lies in the south-eastern corner of the present region, there is only a minor break below the Upper Inferior Oolite (Table 6, col.3). Outside the basinal area, along the marginal belt to the west and south, the Upper Inferior Oolite, in common with the underlying divisions, becomes attenuated. At the same time, the break at the base becomes more important and, locally on the Yeovil 'High', the beds come to rest directly on the Upper Lias.

DUNDRY AND COTSWOLD AREA

Upper Trigonia Grit

Over the Cotswold area, the Upper Trigonia Grit forms the basal member of the Upper Inferior Oolite and consists of 1 to 2.5 m of grey, splintery, shelly ragstone resting on a bored and eroded surface of Lower or Middle Inferior Oolite. Ammonites are rare in the Cotswolds, but brachiopods are abundant and include *Stiphrothyris tumida* and *Acanthothiris spinosa;* the bivalves *Trichites* and *Trigonia* are also common, the latter in the form of casts, which are so numerous as to give the rock its name.

At Dundry, the equivalent beds, where present, are no more than 0.3 m, thick but they are fossiliferous and yield *Rhactorhynchia subtetrahedra*. In the neighbourhood of the Mendips, as at Maes Knoll and Timsbury Sleight, the Upper Trigonia Grit is represented by a thin conglomerate, while it is missing over the Mendip and Vale of Moreton 'axes'.

Dundry Freestone

At Dundry Hill, above the equivalent of the Upper Trigonia Grit, there is a local development of massive limestone, known as the Dundry Freestone. This deposit is only 1 m thick in the eastern part of the hill, but in the western part it reaches 8 m. St Mary Redcliffe and other Bristol churches were built wholly or partly of this freestone, but the quarries are now abandoned, much of the best stone having been long since worked out.

Overlying the Dundry Freestone are coralline beds up to 6 m thick whose top is not seen. These distinctive deposits of crystalline and siliceous rubbly limestone and marl contain *Isastraea* and other corals, as well as echinoderms and brachiopods such as *Zeilleria waltoni* and *Aulacothyris carinata*. Many of the fossils are beekitised, and small irregular wisps and patches of chalcedony investing geodes lined with small quartz crystals are common both in the coralline beds and the upper part of the underlying Freestone.

Upper Coral Bed

Along the main outcrop in the Bath area, as far south as Writhlington and northwards to around Tormarton, coralline limestones very similar to those on Dundry Hill form a distinctive stratum, either resting directly on the Upper Trigonia Grit or separated from it by a metre or so of shelly limestone (? Dundry Freestone equivalent, but see below). In either case, the immediately underlying bed is strongly bored and often oyster covered. The thickness varies from around 0.3 m to locally as much as about 4 m. The upper surface is also strongly bored. Northwards from Tormarton, discontinuous occurrences of apparently similar limestones have been recorded, mainly from the more westerly outcrops in the Wotton-under-Edge – Dursley area.

Doulting Stone – Anabacia Limestone

North of the Mendips and extending into the south Cotswolds, the main mass of the Inferior Oolite above the Upper Trigonia Grit, or Upper Coral Bed if present, comprises white and cream-coloured oolites some 7 to 10 m thick. In good exposures they can be divided into two members, the Doulting Stone below and the Anabacia Limestone above, usually separated by a bored surface. The former consists of massive, more or less current-bedded freestone and the latter, which measures some 3 m or so, is flaggy below and rubbly above. The Anabacia Limestone is so named after the common occurrence in it of the button coral *Chomatoseris* [*Anabacia*] *porpites*, although this form is not confined to these beds.

Clypeus Grit

The name Clypeus Grit is derived from the local profusion of the large sea-urchin *Clypeus sinuatus*, which characterises the formation. Around Stow-on-the-Wold, this fossil is so abundant that when fields are cleaned the heaps of stones are found to be largely made up of damaged specimens. The rock itself is highly distinctive, being almost a pisolite with large yellow granules set in a buff chalky matrix. In the Stroud area and southwards, the echinoid is rarer and the rock has been informally divided into a more marly lower part, which is oolitic rather than pisolitic, a middle member descriptively named the White Oolite Beds, and an upper rubbly part. The formation appears to pass laterally into the Doulting – Anabacia limestones southwards from Horton.

Correlation between The Cotswolds and the area to the South

Ammonite evidence proves that the uppermost parts of the Clypeus Grit and the Anabacia Limestones are of earliest Bathonian age. Also that the remainder of the Clypeus Grit is of *bomfordi* subzonal age and hence, by implication, much of the Anabacia Limestone is too. The Upper Coral Bed at Wotton-under-Edge is considered to be of *truellei* subzonal age and the Upper Trigonia Grit belongs to the *garantiana* Zone, probably the *acris* Subzone. It follows that the Clypeus Grit, for the most part, sits unconformably on the Upper Trigonia Grit and that *truellei* times were mainly marked by nondeposition or erosion in the more northerly areas. On the south side of the Mendips, the lower third of the Doulting Stone appears to be of *garantiana* Zone age, and as there is no evidence of major sedimentary interruption in the beds above, it may be assumed that the overlying beds are largely attributable to the *truellei* Subzone. North of the Mendips, sedimentation during *garantiana* times was much reduced compared to farther south, with relatively thin Upper Trigonia Grit only being represented. Its upper surface represents a nonsequence and the Upper Coral Bed formed as an irregular sheet upon it in the Bath area, but probably only as discontinuous patches farther north. After a further hiatus, probably accompanied by erosion, the spread of detrital and oolitic rocks proceeded slowly northwards reaching over the whole area during later *parkinsoni* Zone times only.

The correlation of the Dundry succession with the main outcrop is uncertain. Recent opinion favours equating the Doulting and Dundry freestones; thus the coralline beds of Dundry would be separate from, and older than the Upper Coral

Bed of Bath. The older view equates the Dundry and Bath coralline beds, thus making the Dundry Freestone older than the Doulting Stone.

AREA SOUTH OF THE COTSWOLDS

Mendip – Doulting – Castle Cary

When traced southwards from Bath the lower part of the Upper Inferior Oolite becomes thinner and eventually disappears (see Figure 36). Only the Doulting Stone and higher beds cross the Mendip Axis, and where these are seen in the deep valleys west of Frome they rest unconformably upon the Carboniferous Limestone of the eastern Mendips.

Magnificent sections showing almost horizontal Upper Inferior Oolite resting on a bored and planed surface of steeply dipping Carboniferous Limestone are exposed at Vallis Vale near Frome. Traces of Lower Lias and Trias occur in some places between the Inferior Oolite and the Carboniferous Limestone. Southwards from Doulting to Batcombe, the surface of the Palaeozoic rocks falls away steeply and higher beds of the Lias appear beneath the Doulting Stone (Figure 36).

Near Doulting, the railway cutting and quarries show typical Doulting Stone consisting of shelly and oolitic limestone and freestone, in which crinoid fragments and oolite grains are embedded in a matrix of crystalline calcite. Considerable lateral variation may be seen in the exposed rock faces and there is much current-bedding. Though not so readily carved as Bath or Ham Hill Stone, Doulting Stone is very durable. It was used in the construction of many fine mediaeval buildings, Wells Cathedral being the most outstanding example.

Together with the overlying Anabacia Limestone, the Upper Inferior Oolite succession measures about 20 m in thickness. Ammonite evidence for the age of the rocks now shows that the uppermost part of the Anabacia Limestone is of earliest Bathonian age, whilst the lowest third of the Doulting Stone, which comprises about 5 m of thinner-bedded more 'raggy' beds than the overlying freestones, is of *garantiana* Zone age. The thickness is exceptional for this zone and may represent a western continuation of the downwarping associated with the Vale of Pewsey Growth Fault (Figure 27). Farther south, the *garantiana* Zone reverts to a more normal thickness of around 1.5 to 3 m. Southwards from Castle Cary, the lower beds pass into a pleasing brown, ferruginous building stone measuring up to 3 m in thickness, locally known as the 'Hadspen Stone'. The overlying limestones are poorly known, but appear to include no economic building stone comparable to the Doulting Stone farther north.

Wincanton – Crewkerne

The Upper Inferior Oolite increases in thickness southwards from the Mere Fault more dramatically than the Lower and Middle subdivisions. The Wincanton Borehole, sited immediately adjacent to the Mere Fault, lies some 6 km to the south-south-east of the Bruton Borehole, which proved a total thickness of 6.8 m of Upper Inferior Oolite. At Wincanton, the total thickness can hardly be less than 45 m and may even approach 50 m. The uppermost 4.7 m comprise oolites, which are presumed to be equivalent to the Anabacia Limestone of farther north, whereas the remainder of the sequence is akin to that in the Dorset – Wessex Basin to the south.

The north-western edge of the main basin is seen in the country around Sherborne and Milborne Port (Table 6, col.3). There, up to 6 m of 'Sherborne Building Stone' is overlain by up to 13 m of rubbly limestone and marl. *Garantiana*

garantiana, Nautilus and 'nests' of *Sphaeroidothyris sphaeroidalis* occur, the last being known to the older quarrymen as 'gooseberries'. Among other interesting fossils, the Sherborne Building Stone has yielded the remains of the dinosaur *Megalosaurus bucklandi*; cones of the conifer *Araucaria cleminshawi* are also thought to have come from these beds.

Above the Sherborne Building Stone and Rubbly Beds lie the Crackment Limestones, a formation which is typically developed in the Milborne Port area. It is composed of pale grey or dirty white, fine-grained limestone with clay partings and is over 11 m thick in its fullest development. The fossils include *Zigzagiceras* and other typical early Bathonian ammonites.

Farther towards the centre of the basin, the Stowell Borehole proved similar facies to those noted at outcrop, but with a total thickness for the Upper Inferior Oolite of about 33 m; the thickness may have increased by a further 4 to 5 m in the Stalbridge Borehole.

To the west of the basin, the thickness variations of the Upper Inferior Oolite closely follow those of the lower beds, with minimum values on the Yeovil 'High' and a thickening towards the western edge of the Inferior Oolite outcrop. The Sherborne Building Stone and the overlying rubbly limestones become more and more attenuated, and at Halfway House are represented by the 'Fossil Bed' and Astarte Bed—two condensed fossiliferous limestones which together total only 0.46 m. Around Yeovil the *garantiana* Zone and *truellei* Subzone are seldom more than 0.2 m thick and can only be recognised with great difficulty.

A similar westerly thinning can be observed in the case of the Crackment Limestones, which are only 1.5 m thick near Yeovil. At Haslebury Mill, near Haslebury Plucknett, it is represented by about 0.1 m of rubbly white limestone and clay with *Morphoceras*. In this condition it is known as the Zigzag Bed and con-stitutes a useful datum in the Crewkerne and Bridport districts.

The thickest recorded Upper Inferior Oolite in the western area is in a quarry near Hinton St. George. Here the *garantiana* Zone consists of hard conglomeratic limestone, of which 0.46 m is seen, overlain by 1.37 m of massive, brown, fer-ruginous limestone similar to the Hadspen Stone of the Castle Cary area. This is overlain by limestones with marly partings attributed to the *parkinsoni* Zone, giving a total exposed thickness of about 5 m.

Sedimentation in the inferior Oolite of the Dorset – south Somerset area

The sedimentation pattern of this and adjacent areas was one of 'swells' or 'highs', on which sedimentation was slow and interrupted, and 'sags' or basins, in which sedimentation was faster and more continuous (Figures 27 and 36). It seems prob-able that the Yeovil 'High' passes southwards into the South Dorset 'High' (Rhys et al., 1982) south of the present region. In the opposite direction, it is possible, though unproven, that the Yeovil 'High' extended north-eastwards to encompass the attenuated successions of the Bruton area and, if so, defined the western edge of the Dorset Basin. The westward thickening of the Inferior Oolite on the western flank of this 'High' may signify that a basin existed to the west of the present out-crop, possibly a northern arm of the Portland Basin (Penn et al., 1980).

Successions in the 'highs' are characterised by conglomerates, erosion planes, evidence of reworking of the sediments, strongly ferruginised coated particles and grains of various sorts, including algal stromatolites, and a general lack of clastic debris. Sediments in the basins are characterised by abundant bioclastic and intra-clastic debris, with or without sand grains and glauconite. Oolites are common to both associations but are ferruginised ('ironshot') in the condensed deposits that characterise the 'highs'. In a paper on limonitic concretions, Gattrall et al. (1972) suggested that the ironshot facies were formed on shallow submarine swells on a marginal shelf, in which there was a minimal supply of sediment that allowed slow concentration of authigenic iron over long periods of time. The lack of clastic debris in the condensed sequences was accounted for by the intervening basins acting as sediment traps. It must be noted, however, that during much of the Middle Inferior Oolite, for instance, the ironshot facies was widespread, even extending across the Cotswolds.

12   Middle Jurassic (Great Oolite Group)

The Great Oolite Group embraces all the formations between the Inferior Oolite and the Kellaways Clay. The pioneer work on this group was done by William Smith in the Bath district in the 1790s and the names Fuller's Earth, Fuller's Earth Rock and Great Oolite derive from this time. Bath is the type area for these rocks which, apart from the Upper Cornbrash and with the addition of the topmost beds of the Inferior Oolite, are assigned to the Bathonian Stage (Table 7).

Although the succession is predominantly clay south of the Mendips, the Fuller's Earth Rock and, more particularly the Forest Marble, form well-marked escarpments. Northwards from the southern environs of Bath, the outcrop broadens with the incoming of thick oolitic limestones, which form the steep scarp and wide uplands of the southern Cotswolds. In the mid and north Cotswolds, where the Inferior Oolite thickens and gives rise to the highest ground, the Great Oolite forms the wide tablelands to the east of the main escarpment that slope gently eastwards and southwards to the low-lying claylands of the Oxford Clay outcrop. The freestones, ragstones and tilestones ('slates') of the group have long been widely used in the Cotswolds for the buildings and dry-stone walls that contribute so much to the beauty of the area.

The correlation of this extremely varied group of rocks is difficult due to the changeability of their sedimentary facies and their faunas, coupled with the rarity of ammonites. Nevertheless, correlation has advanced much in recent years, partly due to new borehole information.

CLASSIFICATION

Arkell and Donovan in an important paper in 1952 on the Fuller's Earth and Great Oolite of the Cotswolds wrote that 'the rocks between the Inferior Oolite and the Cornbrash are the least understood in the Jurassic'. Until then the traditional view, originally put forward by William Smith and developed in detail by Wood-ward in the latter part of the 19th century, was that the Lower Fuller's Earth, Fuller's Earth Rock, Upper Fuller's Earth, Stonesfield Slates, Great Oolite, Bradford Clay and Forest Marble were independent sequential formations. However, as early as 1901 Buckman pointed out that, on the evidence of ammonite faunas, the Fuller's Earth Rock was contemporaneous with the lower part of the Great Oolite of Minchinhampton in Gloucestershire. The stratigraphical implications of this interpretation, however, were not fully appreciated until the investigations by Arkell and Donovan cited above, and by others, notably the British Geological Survey, later.

The name Fuller's Earth has been used since the time of William Smith for the clays that occur between the top of the Inferior Oolite and the base of the Great Oolite or, where this is absent, the Forest Marble. It derives from the occurrence in the Bath area of a bed of commercial fuller's earth within the clay sequence. The more recent extension of the term to include the dominantly limestone successions of the mid-Cotswolds, which are considered to be the lateral equivalents of the clay farther south, is not followed here and the term Fuller's Earth is used in a litho-stratigraphical sense to refer to the clay facies.

The name Great Oolite is used for the main oolite sequence of the Bath area within the Great Oolite Group. The term Great Oolite Limestone, which has been loosely used synonymously with Great Oolite, refers to strata of different ages both in the north Cotswolds and in the Midlands; it is therefore, not used in this account.

The ammonite zones of the Bathonian in Britain are not subdivided, apart from those of *Zigzagiceras zigzag* (three subzones) and *Clydoniceras discus* (two subzones). The *Prohecticoceras retrocostatum* Zone of earlier accounts has been renamed the *Procerites hodsoni* Zone. Unfortunately, ammonites are rare in the Great Oolite Group except in parts of the Dorset Basin, so that even today, after a trickle of discoveries in the last two decades, the limits of the majority of the zones can only be regarded as provisional.

DEPOSITIONAL PATTERN

The Great Oolite Group of the district falls into two main provinces. South of the Bath – Mendip area, and covering Dorset and part of Somerset (Figure 30), the sequence is mainly argillaceous and comprises the Fuller's Earth, the Frome Clay, the Forest Marble and the Cornbrash. Together, these formations thicken southwards into the Wessex Basin where the thickest sequences, in excess of 300 m, lie in a growth faulted belt beyond the margins of the district. Northwards and north-eastwards from this clay-dominated basinal area lies a wide zone covering much of Gloucestershire and Wiltshire, in which most of the clays pass laterally north-eastwards into fine-grained detrital limestones and oolites. These in turn give way to the thinner but varied sequences of Oxfordshire and north Gloucestershire in which micritic limestones form an important part, and which are themselves marginal to the low-lying London Platform landmass east of the borders of the district. Only the overlying Forest Marble and Cornbrash extend across the entire district without substantial facies changes, though the former is notably thicker and more argillaceous in the basinal area.

The successions are broadly interpreted in terms of a relatively deeper-water open sea to the south, which passed northwards and north-eastwards into a shallow marine shelf with tranquil conditions on its deeper water margins and turbulent water farther inshore, where there were shifting carbonate sand banks and shoals. Yet farther inshore, the shoals gave way to shallow lagoonal conditions with evidence of periodic brackish and freshwater influence, and even occasional emergence. The position of the edge of the shelf fluctuated between north and south in Bathonian times; each successive transgression, represented by the clays, penetrated less far north, while the intervening phases, represented for instance by the oolite shoals, migrated progressively farther south (Figures 29, 30).

As the facies belts moved backwards and forwards in response to the relative changes in sea level, they left their mark at any one place by vertical cyclical alternations of sediment. Starting with a transgressive phase, a generalised ideal cycle would comprise: marine clay passing up into fine-grained detrital limestone with or without pisoliths (e.g. Fuller's Earth Rock, Tresham Rock, Twinhoe Beds), passing up into oolites (e.g. Bath Oolite), passing up into micrites (Coppice Limestone, part of the White Limestone) and then into lagoonal and estuarine terrigenous sedimentary rocks (e.g. Sharp's Hill Formation, Hampen Marly Formation, parts of the Forest Marble). This full sequence, however, was rarely achieved.

Comparison of the isopach map of the group (Figure 31) with that for the Inferior Oolite (Figure 27) shows a general resemblance in the eastern Mendip–Wincanton area and in the north-east, where the limit of the Lower Fuller's Earth approximates to the line of the Moreton 'Axis' (Figure 31). West of this 'axis', although the pattern of thickness variation (not shown separately) in the Lower Fuller's Earth in the Cotswolds follows that of the Inferior Oolite, the variations for the Great Oolite Group as a whole are quite different.

STRATIGRAPHY

The Great Oolite Group succession below the Forest Marble is conveniently described according to depositional province.

DORSET–SOMERSET PROVINCE

The general sequence, including the incidence of ammonites, for this province is shown in Figure 30 and Table 7. The Mere Growth Fault appears to have been active during the early and middle parts of the Bathonian, such that the Fuller's Earth thickens dramatically south of the fault.

Lower Fuller's Earth

The Lower Fuller's Earth consists of mudstones with a few thin limestones, and ranges from 35 to 55 m in thickness. The lowest 2 m or so directly overlying the Inferior Oolite comprise two units of widespread occurrence that also extend northwards into the Cotswolds province. At the base is the Fullonicus Limestone, a pale grey, fine-grained limestone which commonly includes ammonites of the uppermost subzone (Table 7) of the *Zigzagiceras zigzag* Zone. This limestone has in times past been confused with rubbly beds at the top of the Inferior Oolite, but is distinguished from the latter by its finer-grained, non-oolitic nature. Its contact with the Inferior Oolite is commonly erosional. The Fullonicus Limestone is overlain by the Knorri Clays, characterised by the distinctive ribbed oyster *Catinula knorri*. The uppermost beds of the Lower Fuller's Earth contain the sickle-shaped *Praeexogyra* [*Liostrea*] *acuminata* in great abundance and have locally been termed the 'Acuminata Beds'.

Fuller's Earth Rock

As typically developed south of the Mendips, the Fuller's Earth Rock consists of rubbly fossiliferous grey limestones with a rich brachiopod and bivalve fauna, and moderately common ammonites. The dominant fossils are brachiopods, of which *Rhynchonelloidella smithi* is a common example. The formation is usually about 4 to 5 m in thickness, but may reach 11 m.

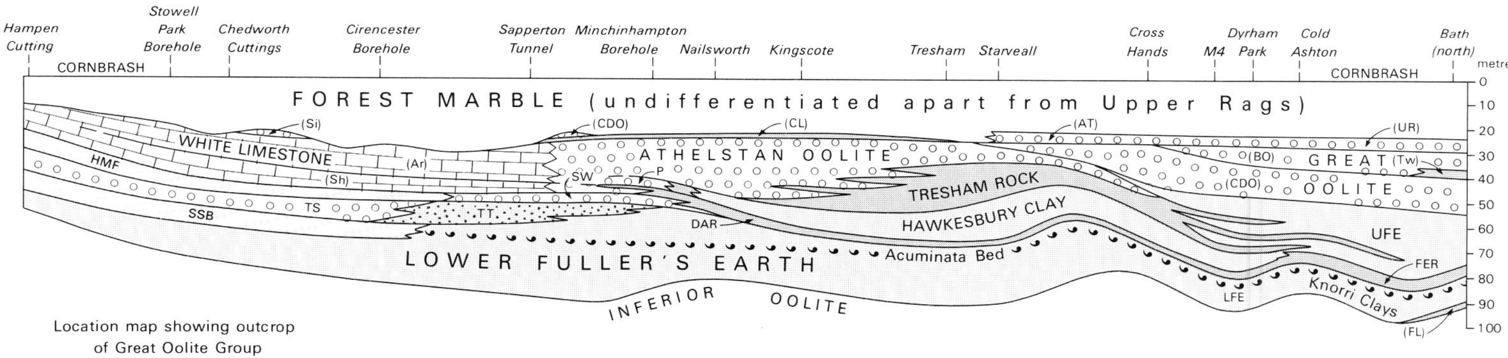

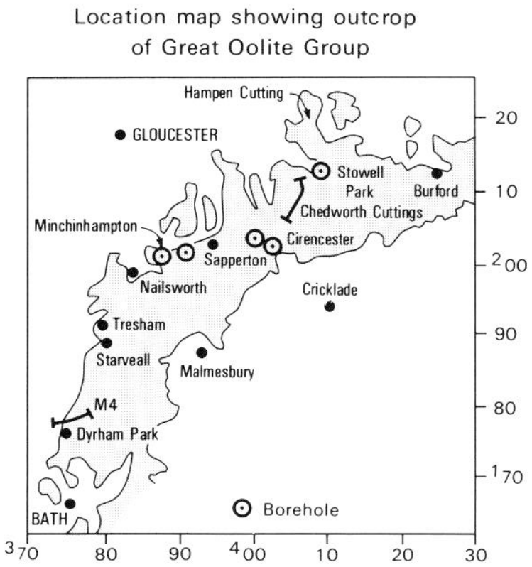

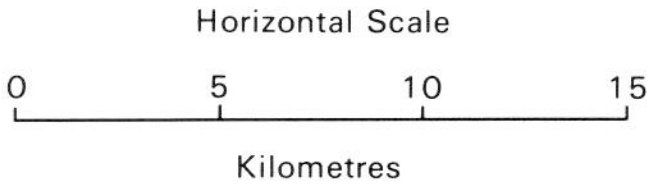

Figure 29   Diagrammatic section to show lateral variation in the Great Oolite Group in the Cotswolds. The section between Kingscote and Cold Ashton is adapted from Cave, 1977, fig. 13.

Abbreviations

(Ar)	Ardley Member	P	'Planking'
(AT)	Acton Turville Beds	(Sh)	Shipton Member
(BO)	Bath Oolite	(Si)	Signet Member
(CDO)	Combe Down Oolite	SSB	Stonesfield Slate Beds
(CL)	Coppice Limestone	SW	'Shelly Beds and Weatherstones'
DAR	Dodington Ash Rock	TS	Taynton Stone
(FEB)	Fuller's Earth Bed	TT	Througham Tilestone
FER	Fuller's Earth Rock	(Tw)	Twinhoe Beds
(FL)	Fullonicus Limestone	UFE	Upper Fuller's Earth
HMF	Hampen Marly Formation	(UR)	Upper Rags
LFE	Lower Fuller's Earth		

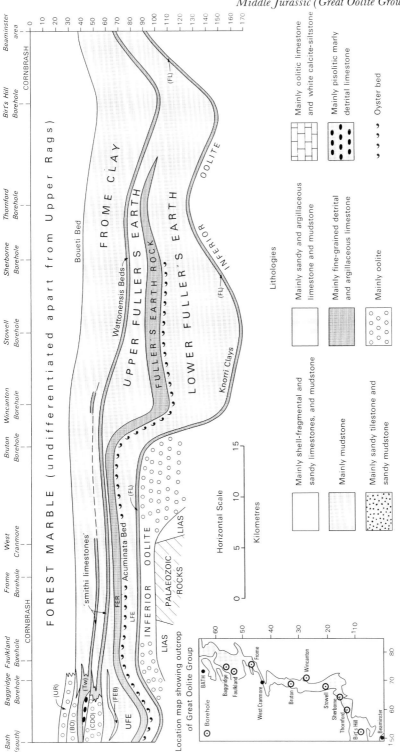

Figure 30 Diagrammatic section to show lateral variation in the Great Oolite Group south of Bath.

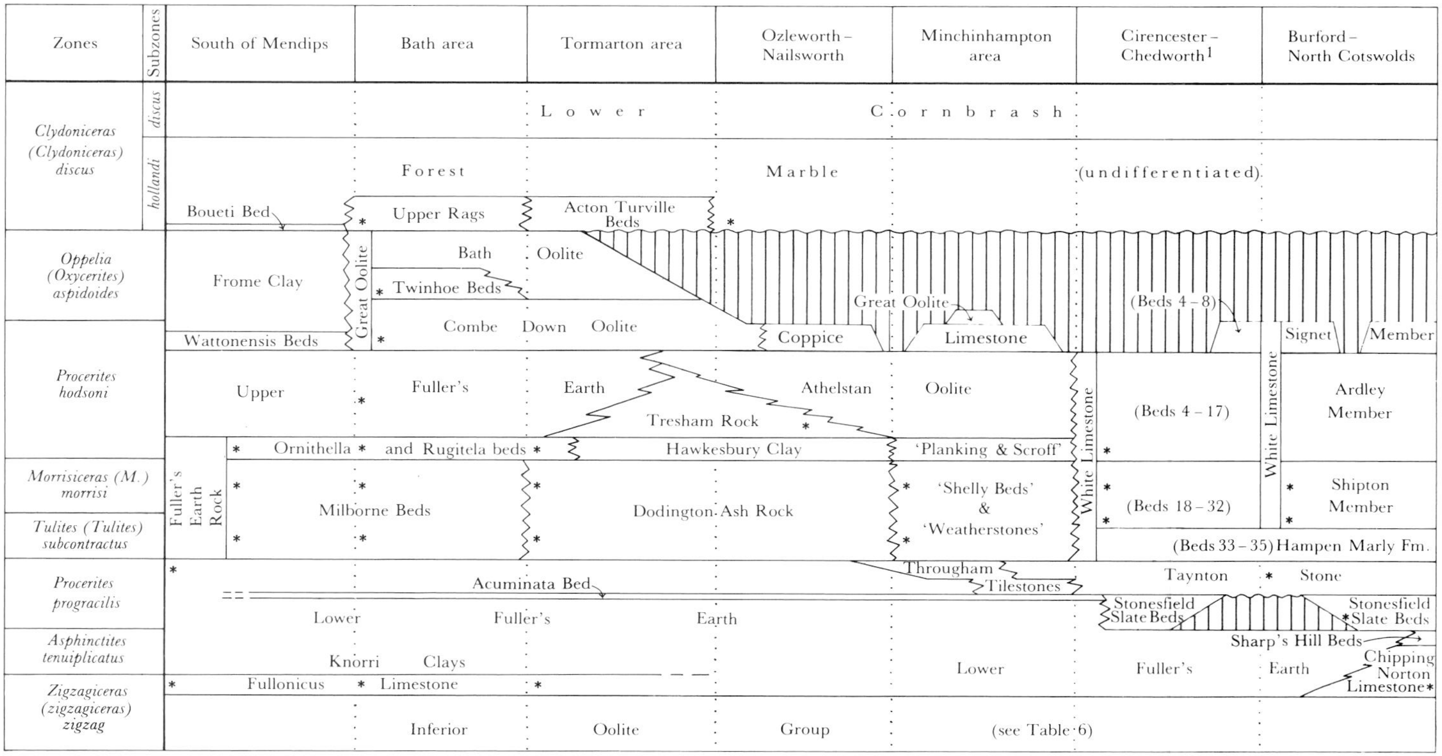

Burford–North Cotswolds | Cirencester–Chedworth[1] | Minchinhampton area | Ozleworth–Nailsworth | Tormarton area | Bath area | South of Mendips

Zones | Subzones

Zones / Subzones (left margin):

Zones	Subzones	
Clydoniceras (Clydoniceras) discus	discus	
	hollandi	
Oppelia (Oxycerites) aspidoides		
Procerites hodsoni		
Morrisiceras (M.) morrisi		
Tulites (Tulites) subcontractus		
Procerites progracilis		
Asphinctites tenuiplicatus		
Zigzagiceras (zigzagiceras) zigzag		

Stratigraphic units (by area):

Cornbrash

Marble

White Limestone — Ardley Member / Signet Member / Shipton Member — (Beds 4–8), (Beds 4–17), (Beds 18–32), (Beds 33–35) Hampen Marly Fm.

(undifferentiated)

Great Oolite / Great Oolite

Coppice Limestone / Athelstan Oolite / 'Planking & Scroff' / 'Shelly Beds' & 'Weatherstones'

Stone / Tayton / Througham Tilestones

Stonesfield Slate Beds / Sharp's Hill Beds / Chipping Norton Limestone

Fuller's Earth

Boueti Bed

Frome Clay

Wattonensis Beds

Upper / Lower

Ornithella and Rugitela beds

Milborne Beds

Acuminata Bed

Fuller's Earth Rock

Fullonicus Limestone

Knorri Clays

Inferior Oolite

Forest Marble — Upper Rags / Bath Oolite / Twinhoe Beds / Combe Down Oolite / Acton Turville Beds

Fuller's Earth

Tresham Rock

Hawkesbury Clay

Dodington Ash Rock

Lower / (see Table 6)

---

[1] The bed numbers in this column refer to the generalized section of the Chedworth–Cirencester cuttings published in the Cirencester (235) Sheet Memoir (Richardson, 1933).

Table 7   Stratigraphy of the Bathonian Stage for the Bristol–Gloucester region

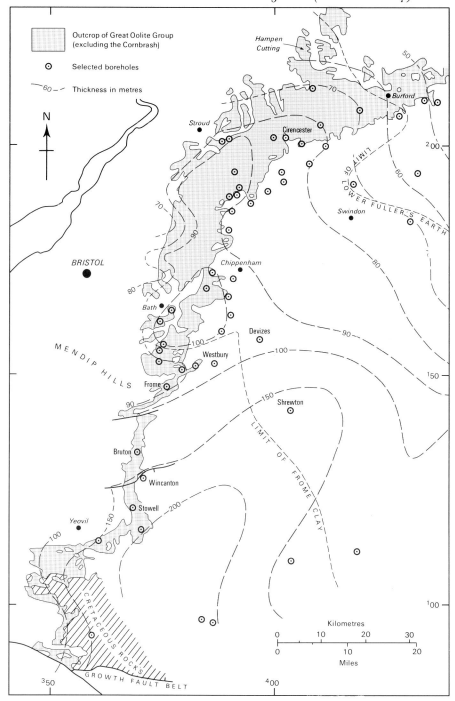

Figure 31 Isopach map of the Great Oolite Group, excluding the Cornbrash. The approximate north-eastern limits of the Lower Fuller's Earth and the Frome Clay are shown.

The lower part of the formation constitutes the Milborne Beds which contain ammonites of the *Morrisiceras morrisi* and *Tulites subcontractus* zones. The upper, more rubbly and argillaceous part, known as the Ornithella Beds, contains an abundance of the brachiopod *Ornithella bathonica* and has yielded ammonites of the early part of the *Procerites hodsoni* Zone. From the vicinity of Stoford southwards to Beaminster the Fuller's Earth Rock is represented only by intermittent limestone nodules.

Upper Fuller's Earth

The Upper Fuller's Earth sequence is not known in detail, but it consists mostly of grey calcareous mudstones with intermittent thin limestones. It is between 8 and 15 m thick in the Sherborne area, but a much greater thickness of 41 m was proved in a borehole at Wincanton.

Frome Clay

The Frome Clay comprises about 30 m of grey calcareous mudstones including, at the base, about 5 m of fossiliferous, nodular, argillaceous limestones, known as the Wattonensis Beds after the abundance in it of the brachiopods *Wattonithyris wattonensis* and *Rhynchonelloidella wattonensis* which, however, are also found in the Fuller's Earth Rock. Other characteristic fossils, many of which are shared with the Fuller's Earth Rock (Rugitela Beds, see below), include the spinose brachiopods *Acanthothiris powerstockensis* and various species of *Rhynchonelloidella,* and the curiously shaped oyster *Lopha marshi.* The absence of ornithellids is considered diagnostic. The beds are a valuable aid for correlation between the south-western part of the basin, where the Fuller's Earth Rock is absent, and the remaining areas, where it is present. Its correlation with the sequences farther north is discussed below.

The Frome Clay is seldom exposed and the stratigraphical ranges of many of the brachiopods and ammonites are therefore uncertain. The main activity of the Mere Growth Fault appears to have ceased after the deposition of the Upper Fuller's Earth; thus the Frome Clay maintains a fairly uniform thickness (Figure 30).

BATH – COTSWOLDS PROVINCE

The northern limit of the Dorset–Somerset province is transitional, with tongues of clay extending northwards for considerable distances into the Bath–Cotswold province. It is, however, convenient to take the Mendip 'Axis' in the Frome area as the dividing line between the provinces because important thickness changes affect the Fuller's Earth here, even though the first major change to the shelf facies, the incoming of the Great Oolite, takes place a short distance to the north (Figure 30).

The influence, if any, of the Mendip 'Axis' on sedimentation in Bathonian times, together with the lateral equivalence of the Great Oolite to the strata farther south, has long been discussed by geologists. Following the primary six-inch survey of the area by the Geological Survey a series of cored boreholes were put down specifically to throw light on these problems. Additionally, a site investigation borehole drilled at Horsecombe Vale on the south side of Bath, in which the Fuller's Earth succession has been divided into 24 distinct sedimentary units, has been proposed as the type section for the Fuller's Earth (Penn et al., 1979).

BATH DISTRICT

The Lower Fuller's Earth and the Fuller's Earth Rock continue across the Mendip 'Axis' into the Bath area virtually unchanged. There is, however, good evidence for strong attenuation over the 'Axis' of the strata between the Fuller's Earth Rock and the horizon of the Wattonensis Beds, though the correlation of the strata on either side of the 'Axis' is disputed, as explained below. The higher Bathonian strata appear to be unaffected by the structure. The current Survey view of correlation across the 'Axis' is given in Table 7 and Figure 30.

Lower Fuller's Earth and Fuller's Earth Rock

The Lower Fuller's Earth comprises 9 units, one of which (Unit 6 of Penn et al., 1979) has been defined as the Acuminata Bed, about 4 m below the base of the Fuller's Earth Rock. This distinctive marker bed, although commonly less than 1.5 m in thickness, persists northwards from some distance south of the Mendips to near Cirencester. The Fuller's Earth Rock comprises two units; the limestones of the lower one form the Milborne Beds, whilst the upper unit consists of the Ornithella Beds overlain by the Rugitela Beds. The type area of the Rugitela Beds is in the eastern Mendips (Sylvester Bradley and Hodson, 1957), where a recently described section gave the complete sequence (Torrens *in* Cope et al., 1980, pp.27–28). Here, the Ornithella Beds of the Fuller's Earth Rock are succeeded by nearly 3 m of mainly rubbly marly limestone, then about 1 m of clay, and finally 0.6 m (seen) of clay interbedded with nodular limestone. The brachiopod and bivalve faunas in both the limestone and the clay with nodules horizon are identical to those of the Wattonensis Beds, and similarly lack ornithellids. The nodular limestone beds also contain an abundant, diverse ammonite fauna of *hodsoni* Zone age. Torrens thus referred all the exposed strata above the Ornithella Beds to the Rugitela Beds and regarded them as equivalent to the Wattonensis Beds of the basinal area to the south and to the Rugitela Beds of the Bath–Tormarton area to the north. Therefore, mudstones assigned by the British Geological Survey to the Frome Clay are regarded by Torrens as being equivalent to the Upper Fuller's Earth. Penn and Wyatt (1979), however, suggest that the lower rubbly limestones be assigned to the Rugitela Beds and the upper nodular limestones to the Wattonensis Beds; the intervening thin clay then corresponds to the 0.5 m of clay, recorded by them as Upper Fuller's Earth in the nearby Frome Borehole (Figure 30). They consider the similar faunas of the two limestones to be facies repetitions, not necessarily of the same age. This alternative interpretation has been substantiated by the results of the Winterborne Kingston Borehole to the south-east of the present region (Rhys et al., 1982).

Upper Fuller's Earth

The Upper Fuller's Earth, which consists mainly of calcareous mudstone with thin limestones in the upper part, ranges from 28 m in thickness at Bath to as little as 0.5 m at Frome to the south. It is divided into thirteen units; the lowest eleven of these units can be resolved into five sedimentary cycles, each of which represents an upward transition from shallow- to deeper-water sedimentation. The shallow-water phase is generally represented by pale, very calcareous silty mudstones or marls dominated by bottom-living forms such as rhynchonellids. These pass up gradually into dark grey to black smooth mudstones representing the deeper-water phase, in which a free-swimming fauna including pectinids becomes important. In boreholes between Bath and Frome the individual units and cycles can be shown to attenuate and overlap each other to rest on the underlying Fuller's Earth Rock as the Mendip 'Axis' is approached.

The commercial Fuller's Earth Bed occurs between Wellow and Bathampton Down. Its vertical distance below the top of the Upper Fuller's Earth varies from 3 to 9 m due mainly to the effects of pre-Great Oolite erosion (see below). It has yielded an ammonite fauna assigned to the upper part of the *hodsoni* Zone. The bed, about 2.5 m in thickness, has long been exploited on the south side of Bath in the vicinity of Combe Hay and Midford, mainly from underground levels driven into the hillside beneath the Great Oolite plateau. Working ceased in 1980. Over 80 per cent of the bed consists of a clay mineral belonging to the smectite group which swells up by absorbing excess water. It also has the property of adsorbing oil; hence the traditional use of fuller's earth in the west country and elsewhere for cleaning the grease from wool and woollen cloth (fulling). Nowadays it has many other industrial uses. The discovery of glass shards and other pyroclastic material in the bed established its volcanic origin, but the source of the ash remains uncertain (Jeans et al., 1977).

Frome Clay

The Frome Clay, as south of the Mendips, consists mainly of calcareous mudstones with a few limestone beds. The sequence passes northwards by lateral passage into the Great Oolite of the shelf facies (Figure 30). Three units have been recognised in borehole cores, each corresponding to a member of the Great Oolite. Persistent shelly limestone beds ('smithi' limestones) occur at the base of the lower two units, of which the one at the base of the formation is considered to be equivalent to the Wattonensis Beds at the base of the Frome Clay south of the Mendips. In a borehole at Frome this bed is separated from the underlying Rugitela Beds by only half a metre of mudstone. South of the 'Axis' this interval expands in conformity with the Fuller's Earth basinal development (Figure 30).

Great Oolite

Bath is the type area for the Great Oolite where it gives rise to the bold scarps and wide plateaux that form such an attractive feature of the district. It is exposed in many quarries (Plate 11B) though few remain in work at the present day. The succession here, fully described by Green and Donovan (1969), is 20 to 30 m thick; in summary it is:

<table>
<tr><td></td><td align="right">*Thickness*
m</td></tr>
</table>

[FOREST MARBLE, UPPER RAGS]

BATH OOLITE
Cream coloured, cross-bedded oolite freestone. Top surface
planed and bored. Base locally sharp   5 to   8.5

TWINHOE BEDS
Buff, compact, fine-grained, shell-detrital limestones with
occasional ironshot pisoliths (Winsley facies), resting on
fossiliferous cream-coloured, pisolitic, rubbly marly limestones
(Freshford facies), with similar but strongly ironshot beds at the
base (Twinhoe Ironshot)   2.5 to 8

COMBE DOWN OOLITE
Cream-coloured, cross-bedded oolite freestone passing down
into shell detrital marly oolitic limestones. Top surface strongly
planed and bored, base often sharp and erosional.   10 to 17.5

The dominant facies of the Great Oolite comprises the massive oolites that have provided the celebrated Bath Stone. This was used by the Romans in the construction of the hot baths; it was also employed in many mediaeval buildings and is still used today. The attractive appearance of the City of Bath owes much to the consistent use of Bath Stone through the ages, more particularly in the 18th century, when the magnificent terraces and crescents were built.

There are three distinct freestone horizons in the Bath area. The lowest occurs in the top half of the Combe Down Oolite. This includes the stone quarried from and formerly mined on Combe Down and Odd Down around Bath itself, and farther east on Box Hill. It is said to be the best weathering variety and can be distinguished by scattered fine-grained shell debris and common calcite stringers (watermarks). The next above is in the Bath Oolite which has been extensively mined in the Westwood – Limpley Stoke – Monkton Farleigh area. It is very pure oolite and is softer than the lower freestone. Lastly, the Upper Rags (Forest Marble) have provided the freestone which has been used in Bradford-upon-Avon from Saxon times onwards. This contains numerous seams of shelly debris that show the cross-bedding to striking effect. During the 19th century the main focus of the Bath stone industry moved eastwards to the Box – Corsham area, following the discovery of thick freestones during the construction of the Box Railway Tunnel. The stone, mainly Bath Oolite, was extensively mined, access being gained by inclines driven through the Forest Marble. Though it hardens on exposure to the air, when newly dug it is readily cut by saws and is ideal for mouldings. The only working mines are at Westwood, Hayes Wood and Monk's Park.

The Twinhoe Beds provide a distinctive datum within the otherwise dominantly oolitic sequence. They are level-bedded and the lower beds are much bioturbated and highly fossiliferous. They appear to have been deposited in deeper, quieter waters than the oolites. The Twinhoe Ironshot includes an important ammonite fauna ascribed to the *aspidoides* Zone. Between Box and Corsham the Twinhoe Beds pass laterally into the Bath Oolite, which is about 18 m thick in the Corsham area.

From its full development on the Twinhoe ridge the Great Oolite has completely disappeared within little more than one kilometre to the south. Mapping and borehole evidence show that the Upper Rags pass into argillaceous Forest Marble facies, while the beds below pass into the Frome Clay (Penn and Wyatt, 1979).

TORMARTON – NAILSWORTH

Lower Fullers Earth

Apart from an increase in thickness, from 16 to 25 m between Tormarton and Nailsworth, the Lower Fuller's Earth continues from the south largely without change in this area. In the southern part of the area, sections between the Inferior Oolite and the Great Oolite were provided by a pipeline trench at Dyrham Park in 1969 and during the construction of the M4 motorway 2 km to the north. These showed that the Acuminata Bed continues from the south (p.133) as a thin, discrete bed about 5 m below the base of the Fuller's Earth Rock. Farther north, due no doubt to increasing estuarine influences in that direction, *Praeexogyra acuminata* also becomes abundant in the overlying strata, which include shelly limestones, and the name 'Acuminata Beds' has been applied to this sequence as in the south (p.127).

Dodington Ash Rock

The M4 motorway sections proved that the Dodington Ash Rock, shown on the survey maps as Cross Hands Rock[1], is equivalent to the Milborne Beds at Bath. The zonal ammonite *Morrisiceras morrisi*, typical of the latter, has been found in the upper part of the rock near its northern limit, about 5 km west of Minchinhampton.

The Dodington Ash Rock is a compact, rather fine-grained limestone with some shell debris. Scattered ferruginous granules and pisoliths are typically present. The thickness varies from 3 to 8 m. North-east of Nympsfield the formation is thought to pass into the 'Shelly Beds and Weatherstones' at Minchinhampton (see below).

Hawkesbury Clay

The Ornithella and Rugitela beds which, at Bath, form the upper part of the Fuller's Earth Rock, are represented north of Tormarton by the Hawkesbury Clay, a dominantly mudstone sequence up to 11 m thick. It thins northwards and passes laterally into the limestone succession at Minchinhampton.

Upper Fuller's Earth

The Upper Fuller's Earth passes laterally northwards from Tormarton into a limestone sequence by the thickening of two main limestone bands at the expense of the intervening clays (Figure 29). The limestones change progressively from argillaceous to finely shell-detrital to oolitic. The lower non-oolitic part of the limestone sequence near Hawkesbury is known as the Tresham Rock, and the upper oolitic phase is called the Athelstan Oolite. The latter progressively replaces the

---

1 The Cross Hands Rock was originally named after an exposure of interbedded thin argillaceous limestones and clays near the Cross Hands Hotel, Old Sodbury. It is now known from unpublished borehole information that these beds occur as a discrete unit near the top of the Lower Fuller's Earth, and the Dodington Ash Rock forms the mapped unit.

Tresham Rock north-eastwards. These oolites were for long confused with those of the younger Great Oolite. The lateral changes are similar to those already noted in the Great Oolite south of Bath, although the complete transition from clay to oolite takes place over a much greater distance (Figure 29).

Great Oolite

The 30 m-thick Great Oolite sequence continues north from Bath as far as Castle Combe. Thereafter, it appears to be progressively overstepped by the Forest Marble (Acton Turville Beds) and is absent from the main outcrop north of Starveall, though farther east it is locally present within a broad belt of oolites extending north-eastwards towards Cirencester. The underlying Athelstan Oolite is generally distinguishable from the Great Oolite by its paler colour, greater purity and eveness of grain.

Along the main outcrop north of Starveall the Coppice Limestone occupies the stratigraphical position of the Great Oolite and is widely overlain by the Forest Marble. Locally, however, outlying patches of Great Oolite are preserved between the two where the Forest Marble overstep has not completely removed it. The Coppice Limestone has been regarded as the top member of the Athelstan Oolite (Cave, 1977), but it could equally well represent a condensed, hardground facies of the Great Oolite beds.

The Coppice Limestone is a distinctive, fawn to cream-coloured, very hard calcite mudstone with a planed, bored and oyster-encrusted surface. The limestone is typical of the micritic hardground beds known as 'Dagham Stone' that occur at intervals in the White Limestone farther north (see below). A characteristic feature of both is the presence of ramifying voids, which it has been suggested are due to the solution of either branching corals or, perhaps more likely, burrow systems. The thickness of the Coppice Limestone is usually 0.3 m to 0.6 m but it may reach 1.5 m. The fauna of bivalves and gastropods indicates that the rock is of primary origin, probably deposited in a sheltered lagoon.

MINCHINHAMPTON

Lower Fuller's Earth

The Lower Fuller's Earth sequence, up to 30 m thick, is much like that farther south, except that another prominent, un-named shell bed rich in *P. acuminata* appears about 2 m below the Acuminata Bed. This shell bed persists north-eastwards into Oxfordshire where it passes into the basal bed of the Stonesfield Slate Beds (see below).

Througham Tilestones

The uppermost beds of the Lower Fuller's Earth, above the Acuminata Bed, pass laterally into the Througham Tilestones, which make their first appearance at Minchinhampton and thicken to a maximum of about 7 m near Bisley. They consist mainly of fissile sandy limestones which were once widely exploited for roofing 'slates'. The tilestones in their turn pass laterally eastwards into the Taynton Stone (see above).

'Shelly Beds and Weatherstones'; Dodington Ash Rock

The Minchinhampton area is important because it marks the transition to the North Cotswold type of succession. The area was made famous by the monographs of Morris and Lycett in the 1850s, which describe the finely preserved mollusca collected from the freestone quarries on Minchinhampton Common. Although the main quarries are closed, quarrying still continues on a reduced scale. The present survey interpretation of the sequence hereabouts is given in Table 7 (col.5).

The 'Shelly Beds and Weatherstones', comprising 7.5 m of cream, cross-bedded oolite freestones with scattered shell debris succeed the Througham Tilestone. They probably represent a north-north-west-trending carbonate bar separating the offshore, open-shelf limestones of the Dodington Ash Rock from the lagoonal limestones of the White Limestone. The overlying 4 m of white, yellowish-weathering, fine-grained, detrital, non-shelly limestone recorded in the old quarries is thought to represent the Dodington Ash Rock, the lower part of which has passed laterally into the 'Shelly Beds and Weatherstones'.

No fewer than 11 species of ammonites ascribed to the *subcontractus* and *morrisi* zones have been recorded from the Minchinhampton area but unfortunately none is adequately localised and their stratigraphical position remains uncertain because their matrix, which is mainly fine grained and nonoolitic, cannot be matched with the quarry descriptions. However, the limited field evidence suggests that the ammonites may have been collected from the 'Shelly Beds and Weatherstones', perhaps in the transition zone to the Dodington Ash Rock at the western margin of Minchinhampton Common.

Athelstan Oolite

The Athelstan Oolite consists mostly of massive, white and creamy-grey, shell-detrital oolite. At the base there is an inpersistent marly oyster bed, the 'Scroff', overlain by about 6 m of coarse, shelly, bedded, sparry oolite known to quarrymen as the 'Planking'. The formation passes eastwards into the Ardley Member of the White Limestone.

CIRENCESTER – NORTH COTSWOLDS

Lower Fuller's Earth

The Lower Fuller's Earth diminishes in thickness due to lateral passage of its upper part into the Stonesfield Slate Beds between Minchinhampton and Cirencester (Figure 29). Its lower part persists for some distance, but is absent north of Condicote and east of the Moreton 'Axis' where it is replaced by the Chipping Norton Limestone and the Sharp's Hill Formation.

Chipping Norton Limestone

The Chipping Norton Limestone, which may include the underlying Hook Norton Limestone of Oxfordshire, reaches a maximum thickness of about 12 m in the Hornsleasow (formerly Snowshill) Quarry, about 4 km west-south-west of Bourton-on-the-Water. This formation exhibits considerable lithological variation. Typically, it consists of buff, hard, rather sandy and splintery oolite, often cross-bedded, containing minute specks of lignite; elsewhere the beds are flaggy and have been worked in the past for flagstones. Decalcification has in some cases reduced it to sand.

Fossils are not common. The bivalve *Plagiostoma cardiiformis* occurs in most exposures, while obscure plant remains and occasional saurian bones of *Megalosaurus* and the crocodile-like *Teleosaurus* and *Steneosaurus* are not infrequently found. Ammonites are rare, but those that occur suggest that the Chipping Norton Limestone falls within the *zigzag* Zone.

Sharp's Hill Formation

Overlying the Chipping Norton Limestone north of Condicote in the north Cotswolds, there are up to 2 m of shelly clays known as the Sharp's Hill Formation, which may include many corals, brachiopods, bivalves and gastropods. They are well exposed at Hornsleasow (formerly Snowshill) Quarry, 4 km west-south-west of Bourton-on-the-Water, where a total of 1.7 m of beds includes a coral bed containing abundant well-preserved compound corals such as *Cyathopora*, *Isastraea* and *Thamnastraea.*

Stonesfield Slate Beds

The Stonesfield Slate Beds comprise varied passage strata, 4 to 10 m thick, between the Lower Fuller's Earth and Taynton Stone north-east of Cirencester. They are most completely seen in the Hampen Cutting of the now disused Cheltenham–Banbury railway line east of Cheltenham. Commonly, the main facies consists of fissile, sandy, shelly, and oolitic limestones comprising the Eyford Member. They were formerly worked for roofing tiles and known as 'Cotswold Slates'. The 'Slates' were obtained from the Slate Bed or 'Pendle', an impersistent bed, 0.6 m or less in thickness, of oolitic sandy limestones near the base of the member. The bed was dug out and exposed to frost action which split the rock along the closely spaced relatively permeable laminae formed of ooliths. The fauna is well known mainly due to the long continued collecting of the slates. Characteristic bivalves include *Gervillella ovata*, *Vaugonia impressa* and *P. acuminata*, but the best-known fossils are vertebrate remains including fragments of pterodactyl, dinosaurs, crocodiles and primitive mammals. The zonal index ammonite *Procerites progracilis* occurs rarely. The thickness varies between 4 m and nearly 10 m.

In the Northleach area a massive freestone member similar to the Taynton Stone and known as the Farmington Freestone is locally developed near the top of the formation.

Taynton Stone

In the Hampen Cutting the Taynton Stone characteristically comprises some 9 m of cream-coloured, cross-bedded oolites with abundant and characteristic seams and wisps of shell detritus. In the Taynton–Burford area it was much worked for freestone and widely used for buildings in Oxford from the Middle Ages onwards; here the thickness reaches up to 12 m. In the area of the Moreton 'Axis' the base of the formation oversteps the Sharp's Hill Formation to rest on the Lower Fuller's Earth. Rare ammonites indicate a *progracilis* zonal age for the Taynton Stone.

Hampen Marly Formation

In the north-eastern extremity of the region, the Hampen Marly Formation, from 5 to 9 m in thickness, consists of grey, green and buff, shelly clays and marls with interbedded buff, grey-hearted silty and sandy limestones, which are locally oolitic. The fauna is dominated by bivalves, with locally conspicuous lumachelles (reefs) of *Liostrea hebridica*; 'nests' of the brachiopod *Kallirhynchia concinna* are common.

Marly limestones become increasingly dominant south-westwards towards Cirencester, where only a few thin marl beds persist. There is a concomitant thinning of the sequence to 3 to 4 m. Beyond Cirencester the formation passes laterally into the Minchinhampton 'Shelly Beds and Weatherstones'.

White Limestone

In recent years subdivision of the White Limestone has been established as far as the eastern edge of the present region (Table 7, col.7; Sumbler, 1984). Correlation and subdivision is by means of widespread and distinctive lithological marker beds, including four main gastropod-bearing, hardground, micritic limestone beds. The latter are distinguished by different species of the high-spired nerineid gastropod *Aphanoptyxis*. The most important section between Cirencester and the Hampen Cutting is provided by the Stony Furlong Cutting, 1.5 km south-east of Chedworth, on the disused Cirencester–Cheltenham railway line. The White Limestone ranges from cross-bedded, 'millet seed', pelloidal oolites to level-bedded, white, splintery calcite mudstone or siltstone with a variable proportion of ooliths and pellets. Clay and marl layers are confined to the lower part of the formation. The most distinctive rocks are the 'Dagham Stone' hardground horizons (see p.137) of which up to five are present; their precise relationship to the hardgrounds recognised in the east remains to be confirmed. However, one persistent bed about midway in the sequence has been identified in the Stony Furlong Cutting and elsewhere as equivalent to the Excavata Bed, which marks the top of the Shipton Member of the Oxford district. The top of the overlying Ardley Member of the Oxford district is defined by the hardground known as the Bladonesis Bed. In the Chedworth Cuttings the beds equivalent to the lower part of the Ardley Member (for long known as the 'Ornithella Beds') are characterised by numerous terebratulid brachiopods, notably *Stiphrothyris* and *Digonella digonoides*.

Beds corresponding to the uppermost member of the eastern Oxfordshire sequence, the Bladon Member, occur between Cirencester and Burford, and are known as the Signet Member. This member is locally absent because of pre-Forest Marble erosion. The rocks mainly consist of rubbly, finely oolitic limestone with a micritic and finely shell-detrital matrix. The lowest beds, which may be marly, commonly contain abundant specimens of the large brachiopod *Epithyris oxonica*. Locally, a fossiliferous coralline facies, known as the Fairford Coral Bed, is developed, in which there is an abundance of well-preserved corals such as *Isastraea* and *Montlivaltia*.

Forest Marble

The term Forest Marble, first used by William Smith, originally referred only to the distinctive shell-detrital limestones lying between the Great Oolite and the Cornbrash, but it was later extended to include all the associated rocks in this position. These form a variable but nevertheless readily recognisable lithological group. Attempts to define the rocks on the basis of the 'Bradford Clay' fauna (see below) have led to confusion in the past. Rare ammonites ascribed to the *hollandi* Subzone have been found in the lower part of the formation.

The dominant facies is argillaceous, but arenaceous and carbonate facies are locally important. In the latter there is complete gradation from pure shelly oolites and calcareous shelly sandstones to sands with doggers. The name 'Hinton Sands' was applied by William Smith to the sandy facies after the type locality of Hinton Charterhouse near Bath where they are about 10 m thick. Like other units in the Forest Marble they are lenticular and impersistent, and the term has little stratigraphical significance. The usual limestone type is hard, blue-hearted, rather sandy, shell-fragmental, flaggy limestone, which weathers to shades of brown and buff; it commonly contains ochreous weathering clay galls, specks of lignite and scattered ooliths. The shell debris is of varied provenance but is dominated by oysters.

The various rock types range from wisps and laminae of a few millimetres to lenticular masses up to 10 m or so in thickness. With the exception of the purer shelly oolites, which tend to be confined to thick massive beds in the more northern areas, every degree of interbedding of the various facies occurs. Small-scale current structures such as ripple marks are widespread and are presumed to have been preserved because of the absence of burrowing organisms. The clays, which are typically unbedded, are grey or green when fresh and weather to whitish and brown respectively.

In Somerset, the Forest Marble consists of a clay series with a middle division of shelly, conspicuously false-bedded limestones, largely made up of oyster fragments, and associated interdigitating sandy facies. From south of the Mendips to the Dorset Coast a thin bed of white-weathering marl known as the Boueti Bed forms the dividing line between the Frome Clay (formerly 'Upper Fuller's Earth') and the Forest Marble. The fossils of the Boueti Bed include abundant *Goniorhynchia boueti* and other brachiopods including *Acanthothiris, Avonothyris, Digonella* and *Dictyothyris*. The ossicles of the 'pear encrinite' *Apiocrinus* are also recorded. The fossils of the Boueti Bed are usually covered with serpulids and the marl composing it sometimes contains pellets of clay apparently derived from the underlying beds. It marks a period when there was pause in sedimentation, probably accompanied by erosion. On the Dorset coast, about 18 m above the Boueti Bed, there is a second fossil band known as the Digona Bed from the abundance of *Digonella digona* in it. There is borehole evidence to suggest that this bed extends northwards into the present district; it is associated with a further pause in sedimentation, for the underlying stratum is strongly burrowed.

The limestones form one of the most prominent of the Jurassic escarpments in south-east Somerset. Starting in the south of the area the Forest Marble outcrop forms the ridge of Birts Hill and runs to Sutton Bingham and Hardington. The greatest thickness of the formation is at Birts Hill where it exceeds 50 m. From here it extends north-east past Sherborne, where it is 40 m or more thick, to Wincanton. At Bowden the limestone was extensively quarried as 'Bowden Marble'. From Wincanton to Frome the Forest Marble is 30 to 40 m or so thick and gives rise to the high ground upon which Frome is built.

The abrupt passage northwards from argillaceous basin facies into shallow-water carbonate shelf facies that characterises the Frome Clay – Great Oolite transition, also affects the lower third of the Forest Marble which passes northwards into the Upper Rags (included on the Bath (265) and Frome (281) sheets with the Great Oolite) just south of Bath. Here the thickness of the Upper Rags ranges between 4.5 and 9 m and the dominant lithology is cream-coloured, shell-detrital, cross-bedded oolite, locally used as a freestone (p.135). A widespread patch reef development at the base is called the Corsham Coral Bed and a more limited development at the top is known as the Bradford Coral Bed. The former is named after the section in the eastern approaches cutting of the Box Tunnel where five patch reefs, together with their associated apron deposits, are well exposed over a distance of 300 m directly underlain by the bored top of the Bath Oolite.

Farther north the Upper Rags were mapped separately as the Acton Turville Beds by tracing field brash from the bored hardground on the top of the Bath Oolite together with coralline debris above it. Northwards from Hawkesbury Upton the Acton Turville Beds are no longer separately distinguished on the Geological Survey maps from the remainder of the Forest Marble.

East of Bath the Bradford-on-Avon district is the type area for the celebrated Bradford Clay. Here the clay is about 3 m thick including a fossil-bed, 0.3 to 0.6 m thick at the base, and is separated from the Upper Rags by about 2 to 3.5 m of rather massive shelly Forest Marble limestone. When well developed the brachiopod fauna is rich and distinctive including such forms as *Digonella digona, Eudesia cardium, Dictyothyris coarctata* and *Epithyris bathonica*. In its type locality in the Canal Quarry, on the south side of Bradford, groups of *Apiocrinus parkinsoni* have been found attached in position of growth to the underlying limestone floor. The fossil assemblage has a number of forms in common with the Boueti Bed. Detailed work in this and surrounding areas has shown, however, that the 'Bradford Clay Bed' is merely one of several impersistent fossiliferous clay beds interbedded with limestones in the lower third or so of the Forest Marble. The Bradford Clay fauna occurs over a few metres of strata and not just as a thin marker band.

Between Hawkesbury Upton and Cirencester there is a considerable but variable development of limestone, mainly cross-bedded, shell-detrital oolite, between the top of the Athelstan Oolite (or White Limestone) and the base of the more typical terrigenous facies of the Forest Marble. Boreholes in the Malmesbury–Kemble area have proved from 9 to 14 m of these beds, and also a variable thickness of the underlying Great Oolite (see p.137).

Defining the base of the Forest Marble within these oolitic limestones has proved difficult. Earlier accounts have taken it at the base of the 'Bradford Clay' and the name 'Kemble Beds' applied to the underlying oolites, but the discontinuous nature of the Bradford Clay has made consistency of application impossible. Survey officers on the Malmesbury (251) and Swindon (252) sheets have regarded the local representative of the Corsham Coral Bed (see above) as the basal member of the Forest Marble. The latter rests on a planed and bored surface which can commonly be recognised even when the coral bed itself is absent. However, even these criteria are locally difficult to apply.

The thickness of the Forest Marble, including the Acton Turville Beds or the Upper Rags, varies between about 26 to 28 m between Bath and Malmesbury, but it thins to about 20 m at Cirencester and more dramatically to about 5 to 8 m in the Burford area. The thickness variations in the latter area are due to the strong channelling of the Forest Marble into the White Limestone. Similar channelling has been noted in the railway cuttings north-east of Cirencester.

Between Cirencester and Fairford the Forest Marble locally furnished tilestones which are known as 'Poulton Slates'.

CORNBRASH

Overlying the Forest Marble is the highly distinctive formation known as the Cornbrash, a term originally applied in Wiltshire to certain stony or brashy soils that are well suited to the growth of cereals. The typical lithology is a brown, fossiliferous non-oolitic, rubbly limestone with abundant shell debris and a marly matrix. Irregular marly partings give the rock a bedded appearance. The Cornbrash, which is often water bearing, frequently overlies a clay formation, and its base is marked by a good feature and by a line of villages associated with springs.

The name Cornbrash was first used in a geological sense by William Smith who acutely observed that the upper part of the formation was different from the lower in the fossils that it contained. This discovery led him to establish the principle of the orderly superposition of strata, distinguished by their fossil content. S S Buckman subsequently published an elaborate faunal analysis of the Cornbrash but this was critically modified in two classic papers by Douglas and Arkell whose account remains the basis for modern stratigraphical descriptions. Unlike the underlying formations, ammonites are sufficiently common to provide a sound basis for zonal subdivision. The Lower Cornbrash is Bathonian in age and the Upper Cornbrash is Callovian; these divisions correspond respectively to the *Clydoniceras (C.) discus* Subzone of the zone of the same name (Table 7) and the *Macrocephalites (M.) macrocephalus* Zone. Four brachiopod faunas can also be recognised, the divisions being as follows:

UPPER CORNBRASH	*Ornithella* [*Micrithyridina*] *lagenalis*
	Ornithella [*Microthyridina*] *siddingtonensis*
LOWER CORNBRASH	*Obovothyris obovata*
	Cererithyris intermedia

According to Douglas and Arkell (1928) the Intermedia Beds are almost invariably present where the junction of the Cornbrash and Forest Marble is seen. The Obovata Beds are probably the most widespread and fossiliferous of the Lower Cornbrash and at the same time show the greatest lithological variation. At some localities *Neocrassina hilpertonensis, Trigonia crucis* and other bivalves occur in such profusion in the upper part of the Obovata Beds as to form a veritable '*Astarte*' – *Trigonia* Bed. It has been recognised by Arkell (1933, p.334), however, that the distribution of the two index brachiopods in the Lower Cornbrash may be ecologically determined and may therefore lack the stratigraphical precision suggested by Douglas and Arkell (see Cope et al., 1980, p.7). The thickness of the Cornbrash ranges from less than a metre to as much as 10 m, but is usually within the range of 1.5 to 6 m. It is about equally divided between the upper and lower divisions. At Frome and Trowbridge, however, on the north-easterly prolongation of the Mendip 'Axis', the Upper Cornbrash is either absent or represented only by a thin shell-lag marl which overlies a karstic surface in the top surface of the Lower Cornbrash.

13 Middle/Upper Jurassic (Kellaways Beds and Oxford Clay)

The Kellaways Beds, named after the hamlet of Kellaways in Wiltshire, have been mapped between Meysey Hampton, 6 km north of Cricklade, and Trowbridge. They are subdivided into the Kellaways Clay and the Kellaways Sand, which forms a discontinuous mappable stratum at the top. The Kellaways Clay consists mainly of silty clay with impersistent thin layers and lenticular beds of sand. It is grey when fresh and at outcrop weathers orange-brown. The Kellaways Sand, where unweathered, is a pale grey, hard calcareous sandstone or sandy fossil-iferous limestone (sometimes known as Kellaways Rock), but at outcrop it is a grit-ty sand. It gives rise to a light sandy loam soil, easily cultivated, and commonly supporting market gardening. The thickness of Kellaways Beds varies from 20 to 30 m, of which the sand commonly accounts for only 2 to 3 m, though it may attain as much as 9 m north-west of Chippenham. Boreholes at Tytherton have provided zonal details of the lower beds. In the southern part of the district the Kellaways Beds have not been separately mapped from the Oxford Clay, though their presence is indicated by lighter soil and more undulating topography, together with evidence from temporary exposures and boreholes.

The youngest Jurassic formation represented in the region is the Oxford Clay, which encompasses most of the Callovian Stage and the greater part of the Lower Oxfordian. The base of the Upper Jurassic is taken at the base of the Oxfordian Stage. Although the formation is on average about 150 m thick it is poorly exposed in the present district. It generally gives rise to heavy clay with a faintly undulating surface channelled by small streams. The Oxford Clay is mainly tough, smooth, greenish grey clay except in the lower part where it is appreciably coarser grained and passes down into the Kellaways Beds.

14 Cretaceous

Cretaceous rocks have a very limited outcrop within the district. To the south-east of Frome high escarpments extending from Longleat to Sturton and the high ground in the neighbourhood of Chard and Crewkwerne are formed by Cretaceous strata. The deposits are described in the British Regional Geology handbooks on 'South-West England' and the 'Hampshire Basin'.

15 Post-Variscan structure and sedimentation

Since the last edition (1948) of this handbook the analysis of Mesozoic sedimentation in England and Wales has greatly advanced under the stimulus of the search for oil and the revolution in geology following the introduction of the theory of plate tectonics. In the classic paper by Godwin-Austin in 1856 and continuing up to Arkell (1933), Mesozoic sedimentation, particularly in the Jurassic, was explained in terms of both major and minor ridges or axes of relative uplift, along which the sedimentation was slow and often interrupted, and intervening troughs of thicker, more continuous sedimentation. These features were said to be controlled by posthumous movement of pre-existing basement structures; for example, axes of uplift were considered to be underlain by anticlinal structures in the basement. In the 1948 edition, the controlling basement structures were described as 'lines of structural weakness' and it was said that 'the seemingly haphazard variations of lithology (are) seen to be intimately related to continuous instability along the axial lines'. The number of axes was augmented, in particular by the addition of the major Bath Axis, which was seen as a southerly manifestation of the old established 'Malvern Line'.

Kent (1949) published a structure contour map of the pre-Permian surface of the whole of England and Wales and delineated the major structural units of the Mesozoic cover, showing them to be on 'essentially different lines from those which controlled pre-Permian sedimentation'. He showed that the axes of uplift were relatively localised features within a framework of large-scale Mesozoic epeirogenic movements controlling sedimentation.

Drilling and geophysical studies, particularly in connection with hydrocarbon exploration, have latterly greatly increased our knowledge and understanding of the structure of England and Wales both offshore and onshore. The data have been incorporated in a series of structure contour and isopach maps recently published for the British Geological Survey (Whittaker, 1985). The results, when combined with modern theories of crustal and lithospheric extension, have provided models for extension-related basin development that are well illustrated in the present region by the Worcester and Wessex basins (Chadwick 1985, 1986). Extension was most obviously achieved by major growth faults, that is faults which were actively moving during sedimentation and which thereby directly affected the pattern of deposition. The locations of such faults appear to have been determined by pre-existing lines of weakness in the basement. Movement on the major basin-controlling growth faults was accompanied by widespread, but smaller-scale, syn-depositional normal faulting. Detailed analysis of seismic sections shows major faulting episodes in Permian – early Triassic, early Liassic and late Jurassic – early Cretaceous times.

MAIN STRUCTURAL ELEMENTS

The main structures of the region are part of major structural units that extend beyond its confines. They constitute massifs, or 'positive' areas, over which Mesozoic sedimentation was much reduced and often interrupted, and 'negative', downwarped basinal areas, in which a considerable thickness of sediment accumulated with little or no interruption.

Figure 32 shows the depth to the top of the pre-Permian Variscan basement and Figure 33 the depth to the top of the Penarth Group. Although some allowance must be made for the initial irregularity of the pre-Permian landscape, comparison of the two maps shows that the main post-Variscan structures were already established before the deposition of the Penarth Group and that subsequent movement resulted in their further development. The broad correspondence of the trend of the Palaeozoic basement structures with that of the Mesozoic structures is also seen.

Owing to considerable erosion of the Mesozoic cover, the contribution of post-Bathonian movement to the various structural elements can only be inferred by reference to immediately adjacent areas and is considered in a concluding section.

Quantock – Cannington Massif

Only the eastern part of the massif is included within the region. It is a Variscan anticlinal structure which formed a positive area in early Mesozoic times; it corresponds approximately to Arkell's North Devon Axis. The Permo-Triassic formations attenuate and/or overlap against it, from all directions. The Lower Lias of west Somerset thins towards the Quantock Hills, although no Jurassic strata now survive in contact with the basement. The south-western side of the Quantocks is formed by the major Cothelstone Fault, a Variscan structure with an estimated dextral horizontal displacement of about 5 km and a vertical downthrow to the south-west of about 2 km. During Permian and early Triassic times it behaved as a growth fault; there is evidence of renewed, smaller post-Lower Liassic movement along it in the Watchet area. The fault continues eastwards across the Wessex Basin as the Cranborne Fault. The nature of the north-eastern boundary of the massif at depth is unknown because of the overlap of the Mercia Mudstone Group onto the Palaeozoic basement, but there may be faulting at depth.

Wessex Basin

The Wessex Basin comprises a series of generally east – west-trending, *en échelon*, fault-controlled sub-basins or half-grabens bounded by growth faults, with intervening 'highs' on which subsidence was less marked. The basin overlies the Variscan fold belt, which extends from the Bristol Channel approaches eastwards to Bridgwater Bay, then eastwards across southern England (Figure 32).

The Central Somerset Basin consists of faulted *en échelon* downfolds. The Permian and the older Triassic strata, proved in Bridgwater Bay, are overlapped towards the margins of the basin by younger Triassic strata, which directly overlie the basement in the marginal areas. To the east, the Pewsey Vale Basin is bounded by the Pewsey Growth Fault and the group of minor faults farther north, which

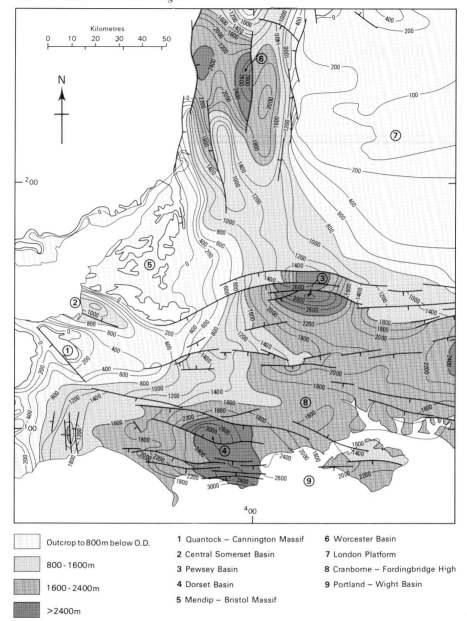

Figure 32   Depth to top of Variscan Basement (after Whittaker, 1985)

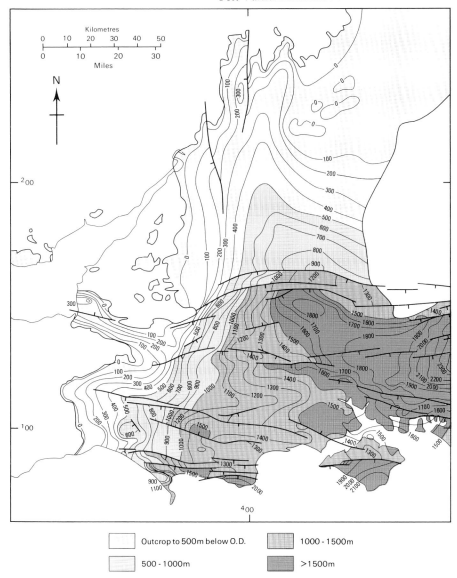

Figure 33 Depth to top of Penarth Group (after Whittaker, 1985). Annotation as for Figure 32. a. Somerton Anticline

overlie the Variscan Front at depth. Farther south, the Mere Growth Fault separates the Bruton 'high' from a minor basin at Mere to the south, and the Coker–Cranborne faults separate the Cranborne–Fordingbridge 'high' from the Dorset Basin to the south (Figure 32). The northerly group of faults, including the Pewsey Fault and the Mere Fault, have been attributed to extensional or normal reactivation of underlying Variscan thrusts identified in the basement by seismic methods (Figure 34). The main depositional centres of the Wessex Basin lie in Dorset and the Weald to the east of the present region.

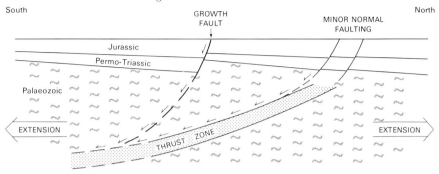

Figure 34   Reactivation and reversal of a Variscan thrust zone to explain Mesozoic growth on the Vale of Pewsey Fault (after Chadwick, Kenolty and Whittaker, 1983, fig. 12).

Mendip – Bristol Massif

This is the positive area lying to the north of the Central Somerset Basin and south-west of the Worcester Basin. Sediments younger than the Lower Jurassic are now largely absent from it due to subsequent erosion.

The southern margin of the massif is formed by the Mendips. The Mendip 'Axis', long figured as one of the major Jurassic axes of uplift, was one of the few that could be directly related to anticlinal structures in the underlying basement. Kent (1949), commenting on the pre-Permian structure contour map, pointed out that 'the Mendip area appears as a shelf-like projection from the Welsh Highlands and the thin development of Jurassic rocks (on the Radstock Shelf) may be regarded as a shoreline condition — an extension of the littoral region of Glamorgan'. The base of the Lower Lias within the massif shows gentle folds with an amplitude of up to 100 m, though an amplitude of 200 m or more is evident in the axial periclinal fold along the Mendips. Although inclination of the beds is partly due to a combination of nearshore depositional slopes and stratal attenuation (i.e. Kent's 'Shoreline condition'), it cannot all be so explained; it is also due to post-Triassic warping and faulting which was greatest over the Mendips themselves.

Worcester (or Severn) Basin

Recent geophysical and drilling information has now defined the form of this north – south-trending basin, in which great thicknesses of Permo-Triassic sediments accumulated in two main depositional areas defined by important growth faults (Figure 32). The faults on the western side of the basin form part of the Malvern Fault Belt (p.65), which was upthrust in Variscan times, after which the movement was reversed in the Permian and Triassic periods. Contrary to many previously stated opinions, the greater part of the basin is now known to be mainly underlain by Precambrian – Tremadocian basement; it thus provides a spectacular example of inversion tectonics (Chadwick and Smith, 1988).

From the south of the Malverns to the south bank of the Severn at Tites Point, the basin-edge is marked by north – south-trending *en échelon* faults which strongly downthrow the Lower Lias and Mercia Mudstone Group strata to the east against

Precambrian and Palaeozoic basement. The geophysical evidence indicates that a steep Triassic–Precambrian/Palaeozoic contact continues beneath the Lower Lias at least as far as the east–west-trending (Variscan) Kingswood Anticline. Attempts by various authors to project the Malvern Line farther south towards the Dorset Coast, mainly on the basis of trends and facies differences in the Mesozoic cover, must be viewed cautiously, not only because of the intense structural overprinting of the basement by the Variscan orogeny but also because of the restricted understanding of east–west facies variation in the cover itself. The eastern limit of the Severn Basin is taken at the Moreton 'Axis', which may also be taken to define the western edge of the London Platform. The basin merges southwards into the Wessex Basin, the dividing line being taken at the change in trend from north–south to a predominantly east–west direction which corresponds at depth to where the Variscan Front thrusts have been identified (cp. Figures 16 and 32).

The Worcester Basin continued to subside throughout Lower Jurassic times, though the depositional centre apparently moved southwards (Figures 32, 33). During deposition of the Inferior Oolite, the southern limit of the basin temporarily retreated northwards (Figure 27).

London Platform

This Mesozoic positive area, which is founded upon relatively weakly deformed Variscan Foreland basement rocks, extends eastwards across Central England and East Anglia into Europe as the London–Brabant Massif. Only its extreme western limit is present in the region, where it terminates at the Vale of Moreton 'Axis'. The latter is a very gently southwards-plunging, north–south-trending asymmetrical anticline. West to east overlap, overstep and attenuation of the strata onto the London Platform occurred in the vicinity of the Vale of Moreton through Triassic into Bathonian times (Figure 36). The earlier literature, rather uncritically, related these features to the present day anticline, but it was later appreciated that they extended eastwards over a much wider area, known as the 'Oxfordshire Shallows'. Borehole evidence has since shown the features to reflect transgressions and regressions from the subsiding Severn Basin in the west across the margin of the London Platform in the east. Structurally, the net effect of the long-continued differential subsidence has been to produce a monoclinal downwarp of considerable amplitude, facing west towards the Worcester Basin in the older Mesozoic rocks (Figure 33). The initiation of this downwarping is generally agreed to be of Permo-Triassic age. The presence of basinward graben-faulting is indicated by the great differences of thickness of Permo-Triassic sediments between the basin and platform; this has been confirmed by geophysical data.

POST-BATHONIAN STRUCTURAL HISTORY

The post-Bathonian structural history of the district remains elusive. Numerous faults, usually normal and with throws of less than 50 m, affect the Mesozoic rocks throughout the district. Most predate the Chalk, but a smaller number postdate it. Recent seismic evidence shows that syndepositional faults may be much more common than had previously been suspected.

At least two important phases of earth movement can be assumed to have affected the structure. First, an important period of erosion took place in earliest

Cretaceous times (the 'Late Cimmerian' Unconformity), following regional uplift and preceding the widespread transgression marked by the Lower Greensand and the Gault. Thus, whereas Lower Cretaceous rocks rest on strata as high as probable Portlandian in the East Bristol Channel Basin, they overlie the Oxford Clay on the eastern margin of the region, near Westbury. It may therefore be conjectured that the region had become a positive area by latest Jurassic or earliest Cretaceous times.

The second important episode occurred during the Miocene and consisted of north–south compressional movements. The most spectacular effects are present to the south of the region in the Dorset coastal area, where major tectonic inversion has long been known. Within the region, where the Cretaceous cover remains, the deformation resulted in reversal of the movement along the Mere and Vale of Pewsey growth faults and in associated anticlinal folding. Away from the Cretaceous outcrops the deformation may be presumed to have enhanced, if not formed the present configuration of such east–west-trending structures as the Somerton Anticline (Figure 33) and the arching of the Mendip Mesozoic cover (see p.150). Re-activation along the line of the Cothelstone Fault (see above) may have occurred during Miocene times, where a post-Liassic reverse fault on the coast at Watchet downthrows 55 m to the north-east and shows a dextral displacement of about 275 m.

16 Quaternary

The Quaternary Period, covering about the last two million years of earth history, is subdivided into the Pleistocene and Holocene. The latter, also known as Recent and corresponding to the Flandrian Stage, accounts for the last 10 000 years and equates approximately to the period since the end of the last glacial episode.

The deposits of the Quaternary Period comprise varied unconsolidated beds, collectively known as 'Drift', including glacial, fluviatile, littoral and estuarine deposits, and a mixed group of periglacial deposits known as 'Head', which grade into colluvium (or hillwash), mudflow and landslip.

Correlation of Quaternary deposits on land is hampered by scarce fossil evidence, considerable lateral variations, general absence of good marker horizons and the complexities of deposition in a glacial environment. Radiocarbon ages can be used, when available, to establish precise chronological dating within the last 40 000 years. The promise of relatively new dating methods such as amino-acid racemisation (AAR) has not yet been realised but, in combination, the various methods offer the prospect of eventually solving the outstanding stratigraphical problems.

The outstanding feature of the Quaternary was the fluctuation in climate between glacial and temperate. The oxygen isotope ratios in shells recovered from deep-sea cores provide good evidence of numerous temperature fluctuations in the oceans throughout the Quaternary, and the relationship between the oxygen isotope stages and the onshore geological successions in north-west Europe is providing an important stimulus for much new research.

The present British classification is largely land-based and relates stages to separate type localities, except for three of the oldest stages. There is broad agreement as to the sequence of events as far back as the last interglacial period (Ipswichian Stage); also an emerging consensus that the great bulk of the pre-Ipswichian glacial sediments in southern Britain can be assigned to one major glaciation (Anglian Stage). There is as yet no generally agreed stratigraphical scheme for the pre-Ipswichian succession.

A postulated sequence of events for the present region is summarised in Table 8, which includes a tentative correlation with oxygen isotope stages, mainly following Bowen et al. (1986, 1988). Several cold and/or glacial periods are represented, although only in the Anglian can an ice-sheet be proved to have entered the region. The deposits of the type area for the Wolstonian stage are now generally agreed to be of Anglian age, but the stage term is retained within quotes because there is evidence in the adjacent Birmingham and South Wales areas of post-Hoxnian glacial deposits, including till, at this stratigraphical level.

The presence of large ice sheets in the vicinity in the late Devensian created widespread periglacial climatic conditions. Local snow-caps were developed and, during the short seasonal thaws, torrents of meltwater carried away rock debris

Table 8   Correlation of the drift deposits of the Bristol–Gloucester region

Oxygen isotope stage	Age (ka) stage	Stages	River terraces Severn	W A*	Bristol Avon & (Frome)	Other drift deposits	Events
1		Flandrian	Alluvium			Estuarine & marine alluvium, peat, etc. of Somerset & other levels	Post-glacial rise in sea level
	— 10 —		1st (Power House) (buried channel deposits)	1st	1st Bristol Avon (inc. buried channel deposits)		
2		Late Devensian	2nd (Worcester)		1st (Strood)	Fan gravels of Cotswolds, Mendips, etc. Cheltenham Sand Cave deposits of Mendips and Wye valley Howe Rock Platform	Glacial diversion of Severn through Ironbridge Gorge at maximum ice advance
			3rd (Main)		2nd (Cairncross)		
	— 25 —					& other low-level wave-cut platforms	
3		Mid	3rd (Leadon tributary	2nd	2nd Bristol Avon	Head & breccia deposits including those overlying raised beach deposits	Upton Warren Interstadial followed by tundra climate
	— 50 —						
4		Early			3rd Bristol Avon (3rd Whitminster)		Mainly tundra climate
	— 75 —						
5a – e		Ipswichian		3rd	?Warm fauna in Bristol Avon 3rd terrace	Raised beach deposits Burtle Beds Ken marine sands	Interglacial
	— 130 —						
6							Cold period
	— 190 —						
7				4th upper		?Raised beach deposits	Temperate period
	— 250 —						
8		'Wolstonian'	4th (Kidderminster)	4th lower		Colluvial deposits on raised beach platforms	Glaciers in Midlands & South Wales
	— 300 —						
9 – 11		Hoxnian (in part)	5th (Bushley Green)	5th/ 6th		Ken River/Yew Tree Farm fluvial deposits Wave-cut platforms up to 14 m +	Temperate with intervening colder stage
	— 430 —						
12		Anglian	Woolridge (fluvioglacial)			'Chalky boulder clay'‡ Paxford gravel‡ Till, fluvioglacial & lake clays in the Vale of Gloucester Campden Tunnel Drift Ken Gravels, Kenpier Till Court Hill gravels & till	Glacial. Bristol Channel glacier, overflow channels in the Bristol area. Lake Harrison in the Midlands
	— 480 —						
13		Cromerian				Westbury-sub-Mendip fissure deposits ?Stretton Sand	Temperate
	— 520 —						
Pre-Cromerian stages						Northern Drift Formation ?High-level deposits in the Bath – Bristol area	Alternating cold/glacial & temperate periods

*Warwick Avon   †(Stroud)?   ‡Moreton Drift

and frozen mud to be deposited in great spreads of fan gravel. Many of the present-day dry valleys in limestone terrain (Plate 2) were probably formed under such conditions; the frozen subsoil did not allow the normal underground percolation of surface water which, as a result, flowed overground and excavated the valleys. Meltwaters from the large ice-fronts fed the ancestors of the present-day rivers, which transported masses of glacial drift, redepositing the material as terraced river gravels.

The sea level varied in proportion to the amount of water locked up in ice. During glacial periods it was appreciably lower than at present, and during warmer periods it was at the present or higher levels. The then prevailing sea levels are reflected in the heights of the ancient river terraces and sea beaches.

EVENTS IN THE LOWER SEVERN AND AVON VALLEYS

The Quaternary history of the lower Severn valley is one in which erosion has dominated over deposition. Nevertheless, sufficient remnants of drift deposits, mainly in terraces of the rivers Severn and Warwickshire Avon and their tributaries, remain to give an indication of the later Pleistocene history of the area.

In pre-Anglian times, the greater part of the present Warwickshire Avon drainage was to the north-east, joining what is now the River Soar. The Avon, at the most, would have been a minor tributary of the Severn. The headwaters of the Severn lay south of Ironbridge, and the lower Severn drainage system was appreciably above present-day sea level. At the height of the Anglian glaciation the lower Severn valley was blocked at least as far south as Gloucester by a glacier deriving from Wales, and the Bristol Channel was occupied by an eastward-moving glacier. Meanwhile, the drainage into the lower 'proto-Soar' valley was blocked by glaciers moving into the area from the north and north-east. These three ice-masses and the main Jurassic escarpments to the south-east confined the drainage into a vast glacial lake that has been named Lake Harrison. Drainage from this lake took place through overflow gaps at about 125 m above OD in the Jurassic escarpment. The Moreton Gap, which occurs in the present district, spilled over into the Evenlode valley, in the Thames catchment.

The oldest terrace deposits of the River Severn represent either fluvioglacial outwash gravels from the retreating glaciers or material redeposited from them. The upper part of the 'proto-Soar' valley was choked with the glacial deposits of the Anglian Stage and, as the ice waned, the River Avon, as we know it, was born by a reversal of the drainage into the Severn basin. The Fifth Terrace of the River Avon has recently yielded a temperate fauna including red deer.

The deposits of the Fourth Terrace of the Avon and, farther upstream to the north of the region, its tributary the Stour at Stratford provide evidence for warm and cold periods between the Anglian and the Ipswichian stages. At Twyning, the Avon Fourth Terrace deposits comprise a basal Jurassic-rich gravel member with a full glacial fauna including mammoth and reindeer, with overlying Triassic-rich sands and gravels including the 'warm' bivalve *Corbicula fluminalis*. The presence of ice-wedge casts in the uppermost part is puzzling and apparently represents a further climatic change (Whitehead, 1988). The Fourth Stour Terrace at Ailstone contains an undoubted interglacial fauna with *C. fluminalis*, and on this basis was long correlated with the Avon Third Terrace in spite of the altitudinal problems involved. It is now known that the Fourth Terrace is both higher and older than the Avon Third Terrace and hence an additional interglacial period is involved (Maddy et al., 1991).

The succeeding Ipswichian (interglacial) Stage is estimated to have lasted from about 130 000 to 75 000 years ago. The Third Terrace gravels of the Avon, which contain a 'warm fauna' including hippopotamus, the straight-tusked elephant *Palaeoloxodon antiquus* and *Corbicula fluminalis*, are assigned to this period. The corresponding Severn terraces appear to have been completely removed during subsequent erosional phases.

The Devensian Stage is divided into two cold parts separated by a somewhat milder period, including the sub-temperate Upton Warren Interstadial. The first of the cold parts corresponds with the Avon Second Terrace, which has yielded a cold fauna including mammoth, woolly rhinoceras, reindeer and various molluscs. After a relatively warm interstadial, the build-up of ice to the north of the region resumed and reached its maximum in the early part of the late Devensian. It resulted in the damming of a large river that had hitherto flowed north-westwards into Cheshire and caused it to divert into the Severn drainage basin, cutting the Ironbridge Gorge in the process. This great increase in the volume of the River Severn led to the intense erosional and depositional activity represented by its Main, Worcester and Power House terraces (Third to First). Recent radiocarbon dating results indicate that this period probably lasted between 25 000 and 10 000 years ago. The former Third Terrace in the main channel of the Severn, which corresponds to the Avon Second Terrace, was replaced by the Main Terrace of later date, following the Ironbridge breakthrough. The Main Terrace and later Severn deposits are distinguished from the earlier gravels by an influx of north-western material derived from the greatly enlarged catchment area, including detritus from the Irish Sea glacier. The great sheets of fan gravel, formed of locally derived Jurassic rocks, that spread downwards from the Cotswolds onto the Severn plain are witness to the rigorous periglacial environment south of the ice-sheets during this period.

Following the building up of the Worcester Terrace during this cold period, there was a fall in the level of the sea to possibly as much as 100 m below OD. The River Severn cut a deep canyon that extended upstream at least as far as Gloucester, and most of the Bristol Channel became dry land.

With the amelioration of the climate at the beginning of Holocene times, sea level rose relatively rapidly until about 6000 years ago, when it stood at about 6 m below OD. Since then, sea level rise has continued at a diminishing rate and sedimentation has reduced, so that in the last 3000 years there has been little change. The climate, as shown by pollen studies, reached an optimum of wet and warm weather around 7500 to 5000 years ago (Pollen Zone VIIa).

Glacial deposits

The glacial deposits of the region are mostly scattered remnants and provide difficult problems of interpretation. The earliest drift deposits are represented by remanié patches of erratic pebbles of quartz, 'Bunter' quartzite and, less abundant, strongly patinated flint lying on the surface of or within fissures in the Cotswold plateau up to a height of 300 m above OD. On the eastern boundary of the present region and in adjacent areas to the east, there are scattered patches of sandy and clayey drift with similar erratics, which are now known collectively as the 'Northern Drift'. The general opinion is that the deposits are heavily decalcified and probably include both tills and the fluviatile deposits derived from them. They predate organic Cromerian deposits in the Oxford area and thus provide evidence for pre-Cromerian glaciation (see summary in Bowen et al., 1986).

High-level plateau deposits in the Bath – Bristol area comprise poorly sorted, loamy gravels with abundant Cretaceous flints and cherts and have been correlated with the 'Northern Drift'.

The Anglian glaciation is better represented in the district. In the Vale of Moreton there is a three-fold sequence. At the base lies the Stretton Sand, a fluviatile, cross-bedded quartz sand, which has yielded a temperate fauna including straight-tusked elephant and red deer. This was formerly dated as Hoxnian in age but now must be considered to be older. The Stretton Sand is similar to the supposedly younger Campden Tunnel Drift (see below), and it has been suggested that the temperate fauna in it is derived from an earlier interglacial deposit. The overlying Paxford Gravel, which comprises local Jurassic limestone material, has yielded mammoth remains and has an irregular erosive contact with the Stretton Sand. At the top, up to several metres of 'Chalky Boulder Clay' with derived 'Bunter' pebbles may be present. Thin red clay is locally present immediately beneath the till, possibly representing a feather-edge remnant of the glacial lake deposits of Lake Harrison (see above).

At the northern end of the Cotswolds, in the gap between Ebrington Hill and Dovers Hill, the Campden Tunnel Drift consists of well-bedded sand and gravel with 'Bunter' pebbles and Welsh igneous rocks, and two beds of red clay with boulders, probably a till. The deposits occupy a glacial overflow channel, up to 23 m deep, caused by the ponding of the Avon and Severn valleys by the Welsh glacier farther downstream (see above).

Evidence in Somerset and Avon, combined with that from South Wales, for an Anglian glacier moving up the Bristol Channel has been accumulating in the last decade or so. The construction of the M5 motorway through the Court Hill Col on the Clevedon – Failand ridge led to the discovery in the bottom of the col of a buried channel, 25 m deep and filled with glacial outwash deposits and till. Drilling has since proved similar drift-filled channels in the Swiss and Tickenham valleys crossing the same ridge. South of the ridge, and rising from beneath the Flandrian alluvium of Kenn Moor, marine, brackish and freshwater interglacial sand and silt overlying red stony and gravelly till and poorly sorted cobbly outwash material were disclosed in drainage trenches and other works. AAR results indicate that whilst the bulk of the interglacial deposits are Ipswichian in age, samples of *Corbicula fluminalis* from fluvial deposits directly overlying the glacial deposits give a much earlier date and suggest that the latter are Anglian in age (Andrews et al., 1984). Similar local occurrences of possible till have been reported beneath the Burtle Beds (see below) of the Somerset levels. In the light of these and other discoveries, the glacial overflow hypothesis of Harmer (1907) for the cutting of the Bristol Avon and Trym gorges has been revived to explain why these rivers cut through hard rock barriers in apparent preference to easier ways through adjacent soft rocks.

Terrace gravels

Fluvioglacial gravels are directly related to an ice-front, unlike river terrace gravels, which are formed within ice-free valleys. The only Severn terrace with claims to be regarded as fluvioglacial is the highest Wooldridge Terrace, known south of Tewkesbury, where its base is about 65 m above the present alluvium surface. This is considered to be the outwash of the Severn valley glacier that was partly responsible for ponding up Lake Harrison (see above).

The main river terraces are numbered upwards according to their height above the alluvium, which is in a series of increasing age. The gradient of most of the terraces on the Severn is greater than that of the alluvium and hence their height above the alluvium decreases downstream; thus the Worcester Terrace (Second) disappears beneath the alluvium at Gloucester.

Generally, the terrace deposits are of sand and gravel, the latter consisting of 'Bunter' quartzite pebbles, flints and rolled fragments of local rocks. The gravel of the older terraces is often extremely coarse in texture. Due to the Ironbridge diversion the Main and later terraces of the River Severn show an influx of Irish Sea glacial erratics such as igneous rocks from Scotland and the Lake District.

The River Frome (Stroud Water) which joins the Severn near Framilode has three terraces, which have been correlated with those of the Severn (Table 8). The gravels consist almost entirely of rolled Jurassic rocks. From the chief of these deposits, the Cairncross Terrace, which is correlated with the Severn Main Terrace, a cold fauna has been obtained: this includes mammoth, woolly rhinoceros and musk ox. The lower part of the terrace merges into fan gravel (see below).

In the coastal regions of Gwent, five gravel terraces are shown on the Survey maps; the first of these is apparently equivalent to the Main Terrace, the second to the Kidderminster Terrace and the fourth to the Bushley Green Terrace of the main Severn sequence.

Terrace gravels occur along the Bristol Avon. At Victoria Pit, Twerton, Bath, the third terrace, with a base at about 42 m above OD and 24 m above the alluvium, is composed mainly of local Jurassic rocks. The fauna includes a mixture of temperate forms, such as the straight-tusked elephant and red deer, and cold forms, such as the mammoth and woolly rhinoceros. The second terrace occurs between Keynsham and Bathampton with a surface ranging from 9 to 15 m above the alluvium. No systematic record of fossils is available. The first terrace gravels fill the valley bottom, including the buried channel, and are aggraded up to about 3 m above the present alluvium surface. The buried channel has been reported upstream as far as Bathampton and the deepest record so far is 17 m below OD at Avonmouth. The mammalian remains reported from the upper surface of the gravel are referable to the Flandrian alluvium. In the absence of more data the correlations shown in Table 8 are probable but unproven.

Head, including Fan Gravel

The widespread Head deposits are a varied group of locally derived, unsorted or poorly sorted materials accumulated by a process of downslope sludging (solifluction), the mobilisation being due to an excess of porewater. Their composition varies according to the upslope parent material, for example mainly silty sand is derived from the Upper Greensand or Upper Lias sands, loamy limestone gravel from Jurassic limestones, and clayey loam or loamy clay from various clay formations. This process appears to have been most active under periglacial conditions during cold periods in the Pleistocene. As larger quantities of water became incorporated into the rock mass, partial sorting occurred and there is a complete gradation downslope into fan gravels and normal waterlain fluviatile deposits. The majority of these deposits date from Devensian times.

Within the present region the most striking manifestations of these processes are the sheets of limestone gravel, both sorted and unsorted, that front the Cotswold scarp between Dursley and Mickleton, and which surround Bredon Hill. They contain a variable proportion of quartz sand believed to be of aeolian (windblown) origin. These sheets grade downwards into the Main and Kidderminster terraces of the River Severn and the First and Second terraces of the River Avon. They are mostly 1 to 2 m thick but may locally attain 4 m. They have yielded Arctic tundra faunas including mammoth, woolly rhinoceros, musk ox, reindeer and lemming. In the Carrent Brook main terrace on the south side of Bredon Hill, radiocarbon dates of 27 000 to 28 000 years confirm a mid-Devensian age for that deposit.

In the neighbourhood of Cheltenham, there are thick deposits of quartzose sand with occasional Jurassic pebble debris, known as the Cheltenham Sand. It is best developed in the valley of the Chelt where, under much of Cheltenham it is 6 m thick and may locally exceed 15 m. Although the constituents are similar to the local fan gravels, their proportions are quite different and it appears to have originated as an aeolian 'cover sand' blown in from the Midlands and redeposited by water action late in the Devensian.

Head and fan gravel deposits front the Mendips, Broadfield Down and elsewhere, where they are typically associated with the lower reaches of gorges and valleys cut in the hills. Close to the Cotswold scarp, masses of unbedded oolitic limestone gravel probably represent ancient screes, whilst the Pleistocene breccias of the Weston–Clevedon district, composed of angular Carboniferous Limestone in a matrix of loamy sand, are probably of the same nature.

Alluvium

Alluvium is more extensive than any other Quaternary deposit. The Somerset Levels, (Plate 12) between the Quantock and Mendip hills, form the second largest

Plate 12   The south face of the Mendips seen across the alluvial flats of the Somerset Levels (A6293).

fenland in England. Other great spreads occur on either side of the River Severn as far north as the Severn Bridge at Aust and beyond.

The alluvial deposits have filled, to mean high-water mark (about 6 m above OD), an extensive buried valley system which was graded to the low sea level prevailing at the end of the Devensian. The thickest succession encountered so far lies to the south-west of Brent Knoll, in the Somerset Levels, where it has been proved to 30 m below OD. The site lies some 7 km inland and it is probable that greater thicknesses are present seaward. The base of the succession is usually marked by a peat layer associated with tree stumps in situ, that are exposed from time to time at low tides along the coast. Radiocarbon measurements show that the age of the basal peat increases with depth below OD, and in the Somerset Levels a maximum of around 8500 years has been recorded at about 20 m below OD. As the sea level rose, the forest died and gave way to swamp, now represented by the basal peat, before being finally overwhelmed by the sea. In the thickest alluvial successions, the basal peat is usually followed by sands, which are succeeded by intertidal, laminated, grey silty clay and fine sand and, finally, up to about OD, by grey clay or peaty clay, representing salt marsh.

Periodic re-invasions by land vegetation along the coastline are represented by thin peat layers around the margins of the levels. About 5500 years ago, however, the widespread clay surface that now lies at about OD was briefly colonised by forests of oak and pine, which were then killed by poor drainage with the establishment of peat mires. Subsequent encroachment of this so-called 'OD peat' by marine clays never extended beyond about 10 km inland from the present coast, leaving raised bogs farther inland. At Shapwick and Meare, these persist to the present day. The inland peat is up to 5 m in thickness and has been exploited, probably since Roman times.

In the remainder of the region the rivers and all but the smallest streams have deposited spreads of loamy alluvial silt.

Raised beach deposits

A locally extensive wave-cut platform at about present sea level occurs at intervals around the present coast. At Brean Down, for instance, the Howe Rock platform, cut in Carboniferous Limestone, surrounds much of the headland and passes beneath late Devensian breccias with an arctic fauna. The inner edge of the platform lies at about OD and the outer edge is more than 6 m below OD. The width of the platform here and elsewhere is such that it seems likely that completion of its cutting may have occupied more than one period of low sea level in Devensian or possibly even earlier times.

Raised beach deposits, and an associated wave-cut platform ranging in height from 9 to 14 m above OD, are present at intervals around the coast between Portishead and Brean Down (Plate 13B), and inland at Weston-in-Gordano and Bleadon. The beach material contains a temperate mollusc fauna shown by AAR tests for Middle Hope, Weston-Super-Mare, to be mainly of Ipswichian age but possibly incorporating somewhat older material (Oxygen Isotope Stage 7). These beach deposits are usually covered with Devensian breccia and solifluction deposits, which may include an Arctic fauna. In the Weston-super-Mare area, colluvial deposits with both cold and warm aspects may intervene between the Ipswichian deposits and the platform. Furthermore, at Holly Lane, Clevedon, quarrying has exposed, beneath Devensian breccias and loams, a wave-cut notch at about 20 m above OD at the junction of a wave-cut platform and an old cliff

Plate 13 Pleistocene geomorphological features near Clevedon, Avon
a. The east Clevedon gap (A10725).
b. Wave-cut platform at about 8 m above OD (A10727).

face. The latter may be correlated with an erosion bench backed by an ancient cliff line in the western Mendips with a base at about 21 m OD.

These various features indicate a complex sequence of events with perhaps three episodes of high sea level, but no agreement exists on the dating of the earlier episodes except that they predate the Devensian glaciation and hence a date earlier than Hoxnian is unlikely.

Burtle Beds   In the vicinity of Middlezoy, and elsewhere in the Somerset Levels, there occurs a series of shelly sands and gravels, known as the Burtle Beds, which are generally considered to represent littoral or sublittoral deposits formed during a period of high sea level. The fossils include common and widely distributed temperate marine shells and, more rarely, the brackish – freshwater *Corbicula fluminalis*. Bones and teeth of elephant, rhinoceros, aurochs and other mammals have also been found and are presumed to have been washed in from the surrounding land areas. Although the bulk of the material is Ipswichian on AAR results, some reworking has occurred because both pre-Ipswichian and Flandrian dates have been locally obtained. Similar deposits have been discovered at Kenn Moor, between Clevedon and Yatton, where they overlie till of presumed Anglian age (see above).

Cave and fissure deposits

There is an extensive literature on the deposits found in the numerous caves and fissures in the Carboniferous Limestone and, to a lesser extent, the Dolomitic Conglomerate of the district. Although the contribution of these deposits to an understanding of the local geology is rarely more than indirect due to their isolated positions, the scattered fossil finds within them bear striking witness to the diversity of the fauna, particularly the larger mammals that inhabited the district in times past.

In recent years, undoubtedly the most interesting discovery has been an extensive fissure deposit uncovered by quarrying operations high above Westbury-sub-Mendip (Bishop, 1982). It has yielded the richest carnivore and small mammal assemblage of any Pleistocene site in Britain; no fewer than eight species were new to Britain. The deposits extend laterally about 160 m and downwards some 20 m from the surface and comprise older bedded silts, sands and gravels and younger unbedded, bone-bearing breccias and conglomerates. The whole complex appears to have filled a large cave system, whose roof has now collapsed.

The older, sandy deposits, washed in from outside, contain a sparse open woodland fauna of which a small *Bison sp.* is the commonest element. This fauna is dated as not later than Cromerian and probably earlier. The later deposits belong to a carnivore lair assemblage dominated by an extinct species of bear but also with many other carnivores including a dhole (*Xenocyon*), a very large lion and an extinct leopard (*Panthera gombazogensis*), the first and last being new to Britain, and their prey. A red-brown earth pocket with remains of small rodents and insectivorous mammals, derived from an owl pellet accumulation, is also present, but its relationship to the breccias is uncertain. The later faunas date from somewhere within the Cromerian to Anglian interval, possibly from a hitherto undescribed interglacial period. A few large flints are thought to be human artefacts and, if so, are the earliest known record of man's presence in Britain. Elsewhere in the district hippopotamus, straight-tusked elephant and narrow-nosed or steppe rhinoceros (*Dicerorhinus hemitoechus*) are recorded from six localities. All are 'warm' forms of uncertain age but were extinct by Devensian times.

The fullest remains are of Devensian age, though many of the largest animals such as cave lion, hyaena, mammoth, woolly rhinoceros and bison apparently disappeared from the area in late Devensian times. Other forms, including reindeer, giant elk, red deer, roe deer, horse, brown bear, lynx, arctic and common fox, hare, lemming and other rodents and small mammals persisted after the Devensian.

Human artefacts and remains have been recorded from about a dozen caves in the district on both sides of the River Severn and are associated with deposits ranging in age from Ipswichian to Holocene. The latest Palaeolithic culture, a native British product known as the Cresswellian, is well represented in the Cheddar area, where it died out at around the beginning of the Holocene. The well-known Cheddar man of Gough's Cave, whose bones are dated at about 9000 years, postdates the Upper Palaeolithic occupation levels but predates an extensive stalagmite layer that is present in most of the district's caves and which appears to be related to a climatic change, presumably an amelioration. It approximately coincides with the rapid postglacial rate of sea-level rise at the beginning of the Flandrian.

17 Economic geology

Coal mining was, for about one hundred years, the most important extractive industry of the district (see p.51). Now the only major activity is quarrying for roadstone, concrete aggregate and, to a much lesser extent, lime. Conforming with twentieth century trends, the former innumerable small workings, using a wide variety of local sources and scattered throughout the district, have now given way to a relatively small number of very large, highly mechanised quarries.

In Victorian times, and until the First World War, the quarrying and mining of Jurassic freestones was widespread and employed large numbers of workers (see p.135). Nowadays, the use of the natural freestone is confined to the highest quality construction and repair work. In the Cotswolds and adjacent areas, however, the Jurassic limestones still provide material for pulverised and reconstituted facing blocks that are widely used where planning controls decree conformity with the original stone buildings.

The digging of clay for brickmaking and other uses, has also greatly diminished as the focus of the industry has moved to the vast Oxford Clay pits of Bedfordshire. Sand and gravel continues to be dug locally, mainly from the river terrace and fan gravel deposits, but also in Gwent from the Quartz Conglomerate (Upper Old Red Sandstone). Peat is dug from the Somerset Levels for horticultural purposes. The annual production of celestite from north of Bristol is around 10 000 tons (see p.83). A long history of fuller's earth mining has now come to an end (see p.134). At Puriton, near Bridgwater, salt (sodium chloride) was obtained in the earlier part of this century by circulating water through boreholes that penetrated the underlying salt beds and then concentrating the brine solution so obtained. Subsequent drilling has indicated the presence of an extensive saltfield in the Central Somerset Basin (see p.81). Gypsum, present as nodular masses and veins in the Blue Anchor Formation, was formerly worked on the foreshore at Watchet. Barite ($BaSO_4$), which is found filling fissures in the Carboniferous Limestone of Cannington Park, near Bridgwater, was also worked on a small scale. Other mineral-based industries have been established within the district from time to time and the most interesting or important occurrences are discussed below.

Roadstone, aggregate, lime

The Carboniferous Limestone accounts for the greatest tonnage of stone sold. There are many large quarries in it on both sides of the River Severn, but the largest concentration is in the eastern Mendips. The limestone is mainly sold as hardcore, roadstone, concrete aggregate, tarmacadam-coated stone and agricultural dust; there is one large lime producer. All the limestone formations are used. The total annual production, which peaked in the early 1970s at about 18 million tons, is of national significance and serves a wide area of southern England.

The Silurian volcanics of the eastern Mendips, the Cromhall Sandstone of north Bristol and the Upper Old Red Sandstone at one locality, though worked on a much smaller scale, are of importance as roadstone because their wearing properties are different from the limestone.

The Blue Lias continues to be worked for hardcore, building stone and lime, principally along the Polden Hills, south of the Mendips. Quarries in the Middle Jurassic limestones of the Cotswolds mainly produce roadstone of various types for local use.

Iron

The iron ores of the Forest of Dean have been worked since Roman times, the ancient outcrop workings being known as 'scowles'. Most of the mining ceased at the beginning of the present century with the approaching exhaustion of the orefield.

The main ores, called 'brown haematites', consist of hydrated ferric oxides with which are included the crystalline form known as goethite. The metallic iron content varies from 15 to 65 per cent. The ore occurs as irregular pockets, lodes and veins, partly replacing limestone in the Carboniferous Limestone (Figure 35). The Crease Limestone and the basal beds of the Whitehead Limestone were the chief repositories: ore was also found in the Drybrook Sandstone, Lower Dolomite and the basal limestone of the Lower Limestone Shale. The orebodies, which decrease with depth, were deposited from descending iron-bearing solutions, the open-jointed limestones being highly susceptible to permeation and metasomatic replacement. The iron carbonates and pyrite of the Coal Measures shales appear to have been the primary source of the iron: weathering under the desert conditions of Permo-Carboniferous times appears to have given rise to an iron-rich surface from which acidic solutions descended to the underlying rocks.

In the Bristol and Somerset area, hematite lodes in the fissured Pennant Sandstone were formerly worked at Iron Acton, Frampton Cotterell, and Temple Cloud, and on a smaller scale elsewhere.

The red and yellow earthy iron oxides known as 'oxide' and 'ochre' have been worked at several localities in the region for the manufacture of pigments. Most recently, Winford and Wick were the chief centres of ochre workings, the material being dug from pockets in the Carboniferous Limestone or from bedded replacement deposits in Triassic rocks adjacent to the limestone.

Lead and zinc ores

The working of lead ores in the Mendips dates from Roman times; the heyday of the industry was probably in the 17th century, but activity continued until the early years of this century. A conservative estimate of the tonnage of lead concentrates produced is in the order of a quarter of a million tons. Serious exploitation of the zinc ores occurred between the beginning of the 17th century and the middle of the 19th century, and the amount produced must have approached that of lead, but no figures are known. Mining apparently ceased with the exhaustion of the readily accessible ore. The depths of the workings do not appear to have exceeded about 100 m.

The main lead ore was galena (PbS) and that of zinc was smithsonite ('calamine', $ZnCO_3$). They occur as veins or fissure-fillings in the Carboniferous Limestone and Dolomitic Conglomerate (Green, 1958), with a gangue of calcite and to a lesser extent, of barite and, rarely, barytocelestite.

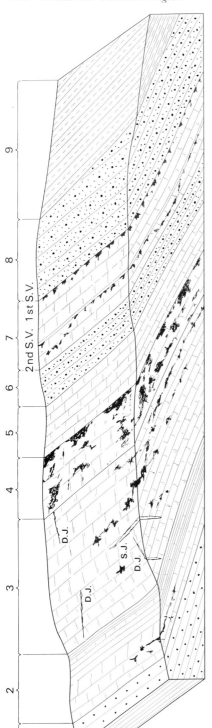

The diagram as a whole does not refer to any particular locality: it assembles features from various parts of the district. Thus:-

In the Drybrook Sandstone: the band of dolomite with the 'First Sandstone Vein' (1st S.V.) and 'Second Sandstone Vein' (2nd S.V.) of iron ore is found only in the south-western part of the district.

The Crease Limestone is generally ore-bearing.

In the Lower Dolomite: joint-veins of ore, illustrated by the ore-bodies marked **D.J.** (dip-joint veins) and **S.J.** (a strike-joint vein) characterize the south-western part of the district. Irregular ore-bodies are well developed in the north-east (Wigpole Syncline).

In the Lower Limestone Shale: deposits of iron ore, represented in the lower or limestone division, are known only in the extreme south.

9. COAL MEASURES
8. Sandstone
7. Dolomite with a course of sandstone } DRYBROOK SANDSTONE
6. Sandstone
5. WHITEHEAD LIMESTONE
4. CREASE-LIMESTONE
3. LOWER DOLOMITE } CARBONIFEROUS LIMESTONE
2. LOWER LIMESTONE SHALE
1. OLD RED SANDSTONE

Figure 35 Block-diagram illustrating the general form and geological distribution of the iron ore deposits (black shading) in the Carboniferous Limestone and Drybrook Sandstone of the Forest of Dean (not drawn to scale).

The main orefield lay in the Central Mendips between Charterhouse and Pen Hill. An important offshoot in the Rowberrow–Shipham area was the main centre of the zinc workings. The unrestored old mine workings are known as 'gruffy grounds' and, though many have been levelled, some can still be seen in the Charterhouse–Lamb Leer cavern area. The ore was taken to be washed and smelted at the 'mineries' near Charterhouse, Priddy and East Harptree, where extensive deposits of the old slags and fines ('slimes') accumulated. The last phase of the industry consisted of reworking these deposits, and all the remaining ruined mine buildings and 'washeries' date from this period.

In addition to the main central Mendips orefield, similar, though more scattered occurrences of ore, mainly galena, have been worked in the Carboniferous Limestone over a very wide area, including much of the remainder of the Mendips, the Weston–Worle ridge and Broadfield Down; also at Pen Park Hole, Brentry and near Westbury-on-Trym, in the Bristol area.

Although there is general agreement that the source of the mineralising fluids is far travelled, it remains uncertain whether juvenile (magmatic) or connate (formation) water, or a combination of both is involved. The Triassic age of the mineralisation, obtained by isotopic dating of the galena, which corresponds to the age of the youngest strongly mineralised rocks (Dolomitic Conglomerate), is no longer accepted as conclusive. It has long been known that small-scale galena-sphalerite–barite-calcite mineralisation affects Penarth Group, Lower Lias and Upper Inferior Oolite rocks in the Mendip–Bristol area, including their secondarily silicified facies (Harptree Beds) within, or close to the main ore-bearing areas of the Mendips and Broadfield Down. The question as to whether the minor and major mineral occurrences are of the same age would be solved if the latter could be linked to the extensive metasomatisation represented by the Harptree Beds. Recent evidence (Stanton, 1981) favours this, and a Middle Jurassic (or later) age for the mineralisation appears increasingly possible.

Manganese

Small quantities of the earthy oxide of manganese, pyrolusite (MnO_2), occur as diffuse pods within iron-hydroxide ore masses in the Dolomitic Conglomerate of the Mendips at East Harptree, Higher Pitts near Ebbor, Croscombe and Wadbury. Much of the ore was used by the potteries for imparting a black colour to the ware, but at Wadbury it was used in the local ironworks to harden steel. It is no longer mined. Of purely scientific interest is the occurrence of a suite of rare minerals, notably oxychlorides of copper and/or lead including mendipite, in cavities within the pyrolusite at Higher Pitts. Comparable occurrences have now been recorded from manganiferous iron ore veins up to 2 m across in Carboniferous Limestone in the eastern Mendips.

Brick, tile and pottery clays

A large brickmaking industry using the wide variety of brick-clays and marls was formerly active within the region. Now fewer than six brickyards remain operational. Household bricks, tiles and pots were made from clays in the Old Red Sandstone, Mercia Mudstone Group, Lias and Alluvium, whilst Coal Measures clays were used for the manufacture of engineering bricks and large pipes. Around Bridgwater, the fine silty alluvium of the River Parret was exploited to make 'Bath Bricks', used for scouring purposes.

Oil shales

Many of the Lower Lias shales are bituminous to some degree. In the early 1920s, sampling of shales in the *bucklandi* Zone on the west Somerset coast showed an oil content of 40 gallons (182 litres) to the volumetric ton (1.15 m^3). Commercial retorts were built at Kilve and produced some hundreds of barrels of oil, but then ceased through lack of financial backing.

Water supply

Large supplies of water are derived from springs and underground sources in a wide variety of formations.

The older Palaeozoic rocks yield only very small amounts of water. Considerable amounts are obtained from the sandstone and the conglomerate at the top of the Old Red Sandstone on the eastern side of the Forest of Dean, but in the Bristol and Somerset coalfields large supplies are not normally present in these rocks.

The Carboniferous Limestone is one of the best sources of water in the region. Large public supplies are derived from perennial springs or underground streams and rivers issuing from the foot of the limestone hills. Such overflow springs are due to the confining of the limestone waters by impermeable Triassic mudstones banked against the limestone and its marginal fringe of Dolomitic Conglomerate. Most of Bristol's water supply, as well as that of Burnham, Weston-super-Mare and many parts of north-west Somerset comes from this source.

Fissured 'pennant sandstone' and other thick sandstones occurring in the Coal Measures also yield much water. Many drowned and abandoned shafts and iron workings have been adapted for water supply, both in the Forest of Dean and in the Bristol Coalfield. Current working of the mines, the draining of the country by adits and pumping from old pits has greatly lowered the water table in these areas. This has led to the drying-up or excessive lowering of the levels of many of the older wells in the Coal Measures and the development of an extensive network of piped supplies fed by central pumping stations.

In the New Red Sandstone both the Dolomitic Conglomerate (mentioned above) and the Triassic sandstones are important aquifers.

The Triassic red mudstones are relatively waterless, or yield hard and saline waters, but sandstones interbedded with the marls yield appreciable supplies in north Somerset, Bristol and north Gloucestershire. Only minor quantities of water are obtained from the White Lias and Blue Lias Limestones, while in the overlying thick clays saline waters, such as those of Cheltenham Spa, are frequently encountered. In central Somerset and the Vale of Gloucester, many small but often unsatisfactory supplies are obtained from patches of head or gravel resting on Mesozoic clays.

Some water supplies are obtained from the Upper Lias sands, for example from the Midford Sands in the Bath district; but on the whole these rocks contain too much silt to yield large constant supplies. The overlying Inferior Oolite and Great Oolite limestones are important aquifers and provide supplies for south-east Somerset and much of the otherwise waterless uplands of the Cotswolds.

Bath – Bristol hot springs

The origin and composition of the Bath thermal waters have been the subject of much scientific curiosity. Notable associated discoveries include the extraction of helium from the exsolved gases by Sir James Dewar, the identification of their radioactivity due to radium by Strutt in 1904 and that due to radon by Munro in 1928. The King's Spring, the principal spring at Bath, emerges through the Lower Lias within a reservoir of Roman construction. It has a temperature of 45°C, which has been constant to within very narrow limits since records began in 1754. Two smaller springs at Bath, the Cross Bath and the Old Royal (or Hetting) Springs, have similar chemical compositions to that of the King's Spring but with temperatures of 41°C and 47°C respectively. Thermal water also occurs at Clifton and Hotwells in the Avon Gorge, Bristol. The Hotwells Spring, which has a temperature of 24°C, discharges from Carboniferous Limestone into river muds just above low tide level.

Systematic investigation of the hot springs has been in progress since 1977 and the results are described in a comprehensive account edited by G A Kellaway (1991). The total yield of the three principal springs at Bath is of the order of 275 000 gals/day (1250 m³/d) and they issue from open fissures in structually complex Lower Carboniferous rocks overlain unconformably by Triassic and Lower Jurassic strata. The floor of the valley at Bath is formed of Lower Lias clay largely obscured by river gravel and alluvium. The fissures postdate the Lower Jurassic rocks and predate the river deposits.

Regional sampling of representative groundwaters from local aquifers show that the Bath thermal water is of unique composition and that the Hotwells water represents a mixture of Bath-type thermal water and shallow Carboniferous Limestone water in the ratio of 1:2.3. The hydrogen and oxygen isotopic compositions of the thermal water demonstrate that it is of meteoric origin. Both the age of the thermal water and the maximum depth at which it is circulating are open to question. Recent estimates of age based on geochemical and radioactivity evidence vary from 'a few hundreds or thousands of years' to 6000–8000 years. The local geothermal gradient is only known within broad limits but it has been calculated, for instance, that within the depth range of the Carboniferous Limestone under the Somerset Coalfield (2700 to 4300 m) a maximum subsurface temperature of 80°C ± 16°C (silica geothermometry) could be attained. A map showing the pre-Roman geomorphology of Bath (Kellaway *in* Cunliffe and Davenport, 1985) indicates the position of streams draining the hot springs in Iron Age times. Archaeological evidence shows that Mesolithic man occupied, at least temporarily, the site of the hot springs, and the position of these remains supports the view that the gravel filling of the King's Spring was already in place by the end of Devensian times. The date of the first emergence of the spring is therefore older than that attributed to the water now rising from them.

References

Most of the references listed below are held in the Library of the British Geological Survey at Keyworth, Nottingham. Copies of the references can be purchased subject to the current copyright legislation.

ALLEN, J R L. 1974. Source rocks of the Lower Old Sandstone: exotic pebbles from the Brownstones, Ross-on-Wye, Hereford and Worcester. *Proceedings of the Geologists' Association*, Vol. 85, 493–510.

— and DINELEY, D L. 1976. The succession of the Lower Old Red Sandstone (Siluro-Devonian along the Ross–Tewkesbury Spur Motorway (M50), Hereford and Worcester. *Geological Journal*, Vol. 11, 1–14.

— and WILLIAMS, B P J. 1979. Interfluvial drainage on Siluro-Devonian alluvial plains in Wales and the Welsh Borders. *Quarterly Journal of the Geological Society of London*, Vol. 136, 361–366.

— — 1981. Sedimentology and stratigraphy of the Townsend Tuff Bed (Lower Old Red Sandstone) in South Wales and the Welsh Borders. *Journal of the Geological Society of London*, Vol. 138, 15–29.

ANDREWS, J T, GILBERTSON, D D, and HAWKINS, A B. 1984. The Pleistocene succession of the Severn Estuary: a revised model based upon amino acid racemization studies. *Journal of the Geological Society of London*, Vol. 141, 967–974.

ARKELL, W J. 1933. *The Jurassic System in Great Britain.* (Clarendon Press: Oxford.)

— and DONOVAN, D T. 1952. The Fuller's Earth of the Cotswolds and its relation to the Great Oolite. *Quarterly Journal of the Geological Society of London*, Vol. 107, 227–253.

BAKER, P G. 1981. Interpretation of the Oolite Marl (Upper Aalenian, Lower Inferior Oolite) of the Cotswolds, England. *Proceedings of the Geologists' Association*, Vol. 92, 169–188.

BISHOP, M J. 1982. The mammal fauna of the early Middle Peistocene cave infill site of Westbury-sub-Mendip. *Palaeontological Association Special Report*, No. 28.

BOSWELL, P G H. 1924. The petrography of the sands of the Upper Lias and Lower Inferior Oolite in the west of England. *Geological Magazine*, Vol. 61, 246–264.

BOWEN, D Q, ROSE, J, McCABE, A M, and SUTHERLAND, D G. 1986. Correlation of Quaternary Glaciations in England, Ireland, Scotland and Wales. *Quaternary Science Reviews*, Vol. 5, 299–340.

— and SYKES, G A. 1988. Correlation of marine events and glaciations on the north-east Atlantic margin. *Philosophical Transactions of the Royal Society of London*, (B), 318, 619–635.

BOYD DAWKINS, W. 1864. On the Rhaetic Beds and White Lias of western and central Somerset; and on the discovery of a new fossil mammal in the Grey Marlstones beneath the bone-bed. *Quarterley Journal of the Geological Society of London*, Vol. 20, 396–412.

BUCKLAND, W, and CONYBEARE, W D. 1824. Observations on the south-western coal district of England. *Transactions of the Geological Society of London*, Vol. 1, Pt. 2, 210–316.

Buckman, S S. 1889.   On the Cotteswold, Midford and Yeovil Sands, and the division between the Lias and the Oolite.   *Quarterly Journal of the Geological Society of London*, Vol. 59, 445–458.

— 1895.   The Bajocian of the mid-Cotteswolds.   *Quarterly Journal of the Geological Society of London*, Vol. 51, 389–462.

— 1901.   Bajocian and contiguous deposits in the north Cotteswolds: the main hillmass.   *Quarterly Journal of the Geological Society of London*, Vol. 57, 126–155.

Butler, M. 1973.   Lower Carboniferous conodont faunas from the eastern Mendips, England.   *Palaeontology*, Vol. 16, 477–517.

Calver, M A. 1969.   Westphalian of Britain.   *6th International Congress of Carboniferous Stratigraphy and Geology, Sheffield, 1967*, Vol. 1, 233–254.

Cave, R. 1977.   Geology of the Malmesbury district.   *Memoir of the Geological Survey of Great Britain.*

— and White, D E. 1971.   The exposures of Ludlow rocks and associated beds at Tites Point and near Newnham, Gloucestershire.   *Geological Journal*, Vol. 7, 239–254.

Chadwick, R A. 1985.   Seismic reflection investigations into the stratigraphy and structural evolution of the Worcester Basin.   *Journal of the Geological Society of London*, Vol. 142, 187–202.

— 1986.   Extension tectonics in the Wessex Basin, southern England.   *Journal of the Geological Society of London*, Vol. 143, 465–488.

— Kenolty, N, and Whittaker, A. 1983.   Crustal structure beneath southern England from deep seismic reflection profiles.   *Journal of the Geological Society of London*, Vol. 140, 893–911.

— and Smith, N J P. 1988.   Evidence of negative structural inversion beneath central England from new seismic reflection data.   *Journal of the Geological Society of London*, Vol. 145, 519–522.

Cope, J C W, Duff, K L, Parsons, C F, Torrens, H S, Wimbledon, W A, and Wright, J K. 1980.   A correlation of Jurassic rocks in the British Isles. Part 2: Middle and Upper Jurassic.   *Special Report of the Geological Society of London*, No. 15.

— Getty, T A, Howarth, M K, Morton, N, and Torrens, H S. 1980.   A correlation of Jurassic rocks in the British Isles. Part 1: Middle and Upper Jurassic.   *Special Report of the Geological Society of London*, No. 14.

Cunliffe, B, and Davenport, P. 1985.   The temple of Sulis Minerva at Bath. Vol. 1 The site.   *Oxford University Committee for Archaeology, Monograph*, No. 7, 4–8.

Curtis, M L K. 1968. The Tremadoc rocks of the Tortworth Inlier, Gloucestershire.   *Proceedings of the Geologists' Association*, Vol. 79, 349–362.

— 1972.   The Silurian rocks of the Tortworth Inlier, Gloucestershire.   *Proceedings of the Geologists' Association*, Vol. 83, 1–35.

Davies, D K. 1969.   Shelf sedimentation: an example from the Jurassic of Britain.   *Journal of Sedimentary Petrology.* Vol. 39, 1344–1370.

Donovan, D T, Horton, A, and Ivimey-Cook, H C. 1979.   The transgression of the Lias over the northern flank of the London Platform.   *Journal of Geological Society of London*, Vol. 136, 165–174.

— and Kellaway, G A. 1984.   Geology of the Bristol district: the Lower Jurassic rocks.   *Memoir of the British Geological Survey.*

Douglas, J A, and Arkell, W J. 1928.   The stratigraphical distribution of the Cornbrash. I The south-western area.   *Quarterly Journal of the Geological Society of London*, Vol. 84, 117–178.

Edmonds, E A, and Williams, B J. 1985.   Geology of the country around Taunton and the Quantock Hills.   *Memoir of the British Geological Survey.*

GATTRALL, L M, JENKYNS, H C, and PARSONS, C F. 1972. Limonitic concretions from the European Jurassic with particular reference to 'snuff boxes' of southern England. *Sedimentology*, Vol. 18, 79 – 103.

GEORGE, T N, JOHNSON, G A L, MITCHELL, M, PRENTICE, J E,RAMSBOTTOM, W H C, SAVASTOPULO, G D, and WILSON, R B. 1976. A correlation of Dinantian rocks in the British Isles. *Special Report of the Geological Society of London*, No. 7.

GODWIN-AUSTEN, R. 1856. On the possible extension of the Coal Measures beneath the south-eastern part of England. *Quarterley Journal of the Geological Society of London*, Vol. 12, 38 – 73.

GREEN, G W. 1958. The central Mendip lead-zinc orefield. *Bulletin of the Geological Survey of Great Britain*, No. 14, 70 – 90.

— and DONOVAN, D T. 1969. The Great Oolite of the Bath area. *Bulletin of the Geological Survey of Great Britain*, No. 30, 1 – 63.

— and WELCH, F B A. 1965. Geology of the country around Wells and Cheddar. *Memoir of the Geological Survey of Great Britain.*

HANCOCK, J M. 1982. Stratigraphy, palaeogeography and structure of the East Mendips Silurian inlier. *Proceedings of the Geologists' Association*, Vol. 93, 247 – 262.

HARMER, F W. 1907. On the origin of certain canon-like valleys associated with lake-like areas of depression. *Quarterly Journal of the Geological Society of London*, Vol. 63, 470 – 514.

HOUSE, M R, RICHARDSON, J B, CHALONER, W G, ALLEN, J R L, HOLLAND, C H, and WESTOLL, T S. 1977. A correlation of the Devonian rocks in the British Isles. *Special Report of the Geological Society of London*, No. 8.

HURST, J M, HANCOCK, N J, and McKERROW, W S. 1978. Wenlock stratigraphy and palaeogeography of Wales and the Welsh borderland. *Proceedings of the Geologists' Association*, Vol. 89, 197 – 226.

JEANS, C V, MERRIMAN, R J, and MITCHELL, J G. 1977. Origin of Middle Jurassic and Lower Cretaceous fuller's earth in England. *Clay Mineralogy*, Vol. 12, 11 – 44.

KELLAWAY, G A. 1967. The Geological Survey Ashton Park Borehole and its bearing on the geology of the Bristol district. *Bulletin of the Geological Survey of Great Britain*, No. 27, 49 – 153.

— 1970. The Upper Coal Measures of south west England compared with those of South Wales and the southern Midlands. *6th International Congress of Carboniferous stratigraphy and geology, Sheffield, 1967*, Vol. 3, 1039 – 1055.

— (editor). 1991. *Hot springs of Bath: investigations of the thermal waters of the Avon Valley.* (Bath: Bath City Council.)

— and HANCOCK, P L. 1983. Structure of the Bristol district, the Forest of Dean, and the Malvern Fault Zone. 88 – 107 in *The Variscan Fold Belt in the British Isles.* HANCOCK, P L (editor). (Bristol: Adam Hilger.)

— and WELCH, F B A. 1948. *British regional geology: Bristol and Gloucester district* (2nd edition). (London: HMSO for Institute of Geological Sciences.)

— and WELCH, F B A. 1955. The Upper Old Red Sandstone and Lower Carboniferous rocks of Bristol and the Mendips compared with those of Chepstow and the Forest of Dean. *Bulletin of the Geological Survey of Great Britain*, No. 9, 1 – 21.

— — 1991. Geology of the Bristol district. *Memoir of the British Geological Survey.*

KENT, P E. 1949. A structure contour map of the buried Pre-Permian rocks of England and Wales. *Proceedings of the Geologists' Association*, Vol. 60, 87 – 104.

LEES, A, and MILLER, J. 1985. Facies variations in Waulsortian buildups, Part 2; Mid-Dinantian buildups from Europe and North America. *Geological Journal*, Vol. 20, 159 – 180.

MADDY, D, KEEN, D H, BRIDGLAND, D R, and GREEN, C P. 1991.   Revised model for the Pleistocene development of the River Avon, Warwickshire.   *Quarterly Journal of the Geological Society of London,* Vol. 148, 473 – 484.

MITCHELL, G F, PENNY, L F, SHOTTON, F W, and WEST, R G. 1973.   A correlation of Quaternary deposits in the British Isles.   *Special Report of the Geological Society of London,* No. 4.

MUDGE, D C. 1978.   Stratigraphy and sedimentation of the Lower Inferior Oolite of the Cotswolds.   *Journal of the Geological Society of London,* Vol. 90, 133 – 152.

MURRAY, J W, and WRIGHT, C A. 1971.   The Carboniferous Limestone of Chipping Sodbury and Wick, Gloucestershire.   *Journal of Geology,* Vol. 7, 255 – 279.

NICKLESS, E F P, BOOTH, S J, and MOSELEY, P N. 1976.   The celestite resources of the area north-east of Bristol with notes on occurrences north and south of the Mendip Hills and in the Vale of Glamorgan.   *Mineral Assessment Report Institute of Geological Sciences,* No. 25.

OWENS, B, RILEY, N J, and CALVER, M A. 1985.   Boundary stratotypes and new stage names for the Lower and Middle Westphalian sequences in Britain.   *10th International Congress of Carboniferous stratigraphy and geology, Madrid, 1983,* 461 – 472.

PENN, I E, DINGWALL, R G, and KNOX, R W O'B. 1980.   The Inferior Oolite (Bajocian) sequence from a borehole in Lyme Bay, Dorset.   *Report of the Institute of Geological Sciences,* No. 79/3.

— and WYATT, R J. 1979.   The stratigraphy and correlation of the Bathonian strata in the Bath – Frome area.   *Report of the Institute of Geological Sciences,* No. 78/22, 23 – 88.

— — and MERRIMAN, R J. 1979.   A proposed type-section for the Fuller's Earth (Bathonian), based on the Horsecombe Vale No. 15 Borehole, near Bath, with details of contiguous strata.   *Report of the Institute of Geological Sciences,* No. 78/22, 1 – 22.

RAMSBOTTOM, W H C. 1973.   Transgressions and regressions in the Dinantian: a new synthesis of British Dinantian stratigraphy.   *Proceedings of the Yorkshire Geological Society,* Vol. 39, 567 – 607.

— CALVER, M A, EAGAR, R M C, HODSON, F, HOLLIDAY, D W, STUBBLEFIELD, C J, and WILSON, R B. 1978.   A correlation of the Silesian rocks in the British Isles.   *Special Report of the Geological Society of London,* No. 10.

REYNOLDS, S H. 1921.   The lithological succession of the Carboniferous Limestone (Avonian) of the Avon Gorge section at Clifton.   *Quarterly Journal of the Geological Society of London,* Vol. 77, 213 – 245.

— and VAUGHAN, A. 1911.   Faunal and lithological sequence in the Carboniferous Limestone Series (Avonian) of Burrington Combe (Somerset).   *Quarterly Journal of the Geological Society of London,* Vol. 67, 342 – 392.

RICHARDSON, L. 1929.   Geology of the country around Moreton-in-Marsh.   *Memoir of the Geological Society of Great Britain.*

RHYS, G H, LOTT, G K, and CALVER, M A (editors). 1982.   The Winterborne Kingston borehole, Dorset, England.   *Report of the Institute of Geological Sciences,* No. 81/3.

SMITH, A H V, and BUTTERWORTH, M A. 1967.   Miospores in the coal seams of the Carboniferous of Great Britain.   *Special Paper in Palaeontology,* No. 1, 1 – 324.

SMITH, S, and STUBBLEFIELD, C J. 1933.   On the occurrence of Tremadoc Shales in the Tortworth inlier (Gloucestershire), with notes on the fossils.   *Quarterly Journal of the Geological Society of London,* Vol. 89, 357 – 378.

STANTON, W I. 1981.   Further field evidence of the age and origin of the lead-zinc-silica mineralisation of the Mendip region.   *Proceedings of the Bristol Naturalists' Society,* Vol. 41, 25 – 34.

SUMBLER, M G. 1984.   The stratigraphy of the Bathonian White Limestone and Forest Marble of Oxfordshire.   *Proceedings of the Geologists' Association,* Vol. 95, 51 – 64.

SYLVESTER-BRADLEY, P C, and HODSON, F. 1957.   The Fuller's Earth of Whatley, Somerset.   *Geological Magazine*, Vol. 94, 312–322.

VAUGHAN, A. 1905.   The palaeontological sequence in the Carboniferous Limestone of the Bristol area.   *Quarterly Journal of the Geological Society of London*, Vol. 61, 181–307.

WALLIS, F S. 1927.   The Old Red Sandstone of the Bristol District.   *Quarterly Journal of the Geological Society of London*,   Vol. 83, 760–787.

WARRINGTON, G, AUDLEY-CHARLES, M G, ELLIOTT, R E, EVANS, W B, IVIMEY-COOK, H C, KENT, P E, ROBINSON, P L, SHOTTON, F W, and TAYLOR, F M. 1980.   A correlation of Triassic Rocks in the British Isles.   *Special Report of the Geological Society of London*, No. 13.

WELCH, F B A. 1933.   The geological structure of the eastern Mendips.   *Quarterly Journal of the Geological Society of London*, Vol. 89, 14–52.

— and TROTTER, F M. 1961.   Geology of the country around Monmouth and Chepstow.   *Memoir of the Geological Survey of Great Britain.*

WHITTAKER, A. 1985 (editor).   *Atlas of onshore sedimentary basins in England and Wales*: Post-Carboniferous tectonics and stratigraphy.   (Blackie: Glasgow and London.)

— and GREEN, G W. 1983.   Geology of the country around Weston-super-Mare. *Memoir of the Geological Survey of Great Britain.*

— and SCRIVENER, R C. 1982.   The Knap Farm Borehole at Cannington Park, Somerset, *Report of the Institute of Geological Sciences*, No. 82/5, 1–7.

WILLIAMS, B J, and WHITTAKER, A. 1974.   Geology of the country around Stratford-upon-Avon and Evesham.   *Memoir of the Geological Survey of Great Britain.*

WILLIAMS, G D, and CHAPMAN, T J. 1986.   The Bristol–Mendip foreland thrust belt.   *Journal of the Geological Society of London*, Vol. 143, 63–73.

WILSON, D, DAVIES, J R, and WATERS, R A. 1988.   Structural controls on Upper Palaeozoic sedimentation in south-east Wales.   *Journal of the Geological Society of London*, Vol. 145, 901–914.

WILSON, V, WELCH, F B A, ROBBIE, J R, and GREEN, G W. 1958.   Geology of the country around Bridport and Yeovil.   *Memoir of the Geological Society of Great Britain.*

WORSSAM, B C, ELLISON, R A, and MOORLOCK, B S P. 1989.   Geology of the country around Tewkesbury.   *Memoir of the Geological Society of Great Britain.*

Index

Printed in the United Kingdom for HMSO
Dd 0291141 2/93 C50 531/3 12521